John Sh

UNITED STATE
AND COAST GUA
IN NAVAL WARFARE,

Table of Contents

Acknowledgments

The motivation to write this book on 19th century naval warfare came from many sources. One of them came as a result of my membership in the Civil War Round Table of Rochester (RCWRT), Minnesota. Richard G. Krom and his spouse, program director Sharon Krom, administer the RCWRT, and Sharon contributed to Richard's book, *The First Minnesota, Second to None*. (The book was published in 2010 by the author and produced by Bang Printing in Brainerd, Minnesota; the couple's son, James Krom, the proprietor of Natural Images in Rochester, created the cover illustration.) *First Minnesota* is based on letters home written by Edward H. Bassett, a Civil War soldier in the First Minnesota Regiment and Richard Krom's great-grandfather.

Troop ships transported Bassett to and from his home and into the maritime domains of the Potomac, James, and York rivers, Chesapeake Bay, and the coastal Atlantic. As the author of books on U.S. revenue cutter and U.S. Coast Guard history, I was drawn to the maritime experiences of the Minnesota soldiers as they traversed rivers and waterways on civilian steamships, military and naval transports, and boats in a variety of logistical and tactical missions. Edward Bassett's letters and Krom's narrative segues described pontoon bridge building across strategic rivers. Bassett's accounts of his travels included references to the paddle-wheel steamers SS *Atlantic,* SS *Baltic,* and SS *Empire City* and the wood-hulled and iron-clad U.S. Navy ships USS *Monitor,* USS *Minnesota,* USS *Keokuk,* and USS *Essex.*

My subsequent research on the age of sail and coal-fueled, steam-powered ships and the transition from wooden-hulled to iron combat vessels led to this book and the joint missions of the U.S. Navy and the U.S. Revenue Cutter Service and U.S. Coast Guard throughout the 19th century and into the 20th. The wars that occurred in that time frame include the Quasi-War with France (1800), the War of 1812–1815, the Seminole Wars of the 1830s and after, naval missions against piracy and slave ships, the U.S.–Mexican War (1846–1848), the Civil War (1861–1865), the Spanish-American War (1898), and the entry in 1917 of the United States into the Great War of 1914–1918 (World War I).

I have been greatly influenced by David Howard Allen, a retired IBM electrical engineer, military history scholar, gamer, lecturer, and supporter of veterans and veteran organizations. Dave has taken course work from American Military University and has supported and encouraged me in my writing, contributed greatly to my knowledge of military history and operations, and attended my oral presentations and classes locally and regionally, including the National World War II Museum in New Orleans, Louisiana, and the U.S. Navy Memorial in Washington, D.C.

David Allen is the illustrator and cartographer for this book; he has graphic design experience in technical papers, is the author or coauthor of 20 published papers, and the inventor or co-inventor of numerous patents. Dave has written extensively about military history, including a published paper, "Twilight of the Luftwaffe," about the rise and fall of the German military machine.

David Allen's respect and support for the military was inspired by his father, Lieutenant Colonel John Charles Allen (U.S. Army), and three brothers who served in the military. Colonel Allen served as an enlisted man in the U.S. Army Air Force in World War II, then the National Guard, and as a commissioned officer in the Korean War (1952–1953). Colonel Allen graduated from the U.S. Command and General Staff College, served on the general staff at the Pentagon in Washington, D.C., and continued his military career until 1968. LTC Allen was buried in Arlington National Cemetery in 2016.

Medical issues prevented David Allen from directly serving in the military and naval forces of the United States, but his siblings served. David's brother John Charles Allen, Jr., was a midshipman first class at the U.S. Naval Academy until his tragic death in an automobile accident. Jeffrey Lewis Allen graduated from the U.S. Air Force Academy and served as a commissioned officer for several years. James Edward Allen graduated from the U.S. Air Force Reserve Officer Training Corps program at the University of Kansas and served as a commissioned officer.

Captain Joseph Connell served in Vietnam as a U.S. Air Force helicopter rescue pilot. His friendship, service, and responsibilities taught me much about command, courage and leadership responsibilities, and he further enriched me with his research on military history. After a sales and administrative career at IBM, Joe immersed himself in veterans affairs, historical research, air shows, and piloting his self-built aircraft. His knowledge and respect for the Coast Guard was inspirational. Captain Connell was familiar with the U.S. Coast Guard in Vietnam and the aviators who flew helicopters on search and rescue (SAR) missions. Mr. Connell supported me in my professional activities and in our joint historical research projects. Joe participated in the historic Veteran Honor Flights to Washington, D.C., and has given presentations on military issues and history.

Master Technician Senior Chief Tina M. Claflin, U.S. Coast Guard (Ret.), inspired and supported my writing and speaking. Senior Chief Claflin's career included sea duty on the U.S. Coast Guard Academy training sailing ship USCGC *Eagle* (WIX-327) and other cutters, leadership in Coast Guard women's support programs. She was among other distinguished female members of the military services honored by the Congressional Women's Caucus. Senior Chief Claflin, a Rhinelander, Wisconsin, native, enriched my presentations at the U.S. Navy Memorial in Washington, D.C., and I have encouraged Chief Claflin's writing and research.

The staff at the History Center of Olmsted County in Rochester, Minnesota, has provided the staff and facilities for historical presentations and research. Program coordinator Aaron Saterdalen has encouraged this author by providing speaking and research opportunities, as has HCOC curator Daniel Nowakowski.

It has been an enriching experience to speak to, and learn from, the dedicated members of the Hiawatha Valley Civil War Round Table in Winona, Minnesota, at the invitation of Round Table director president Anthony Straseske.

Robert Fuhrman, the executive director of the Richard I. Bong Veteran's History Center in Superior, Wisconsin, has provided me with several opportunities to speak at the acclaimed museum as well as the experience of learning about military history from him and his professional staff.

My cousin, Lt. Col. George Ostrom, United States Air Force (Ret.), has inspired me with his humor and leadership skills, his knowledge and experience in the military, intelligence, and defense communities, and his presence and support at my presentations at the United States Navy Memorial in Washington, D.C.

Mark Weber, the eclectic curator at the magnificent United States Navy Memorial and its museum in Washington, D.C., and his staff have been gracious and inspirational in providing me with the opportunity and honor of speaking at that distinguished institution.

My wife, Mary Lamal Ostrom, a former elementary school teacher, has supported my writing, speaking and travels in ways that made this book possible.

No author works alone. The list of associates and friends enumerated above, and the sources indicated in the narrative, illustrations, and bibliography of this book, illustrated the truth of that observation.

Preface

The United States Coast Guard was founded in 1790 during the administration of President George Washington. It was initially called the U.S. Revenue Marine (USRM) and later the U.S. Revenue Cutter Service (USRCS). Congress placed the naval service in the Treasury Department under Secretary Alexander Hamilton. In 1915, during the administration of President Woodrow Wilson, the Revenue Cutter Service was combined with the U.S. Life-Saving Service to form the United States Coast Guard. The USCG would be moved from the Treasury Department into the Department of Transportation in 1967 and to the Department of Homeland Security in 2003.

This book will emphasize the domestic and wartime missions of the USRM/USRC/USCG from 1790 to 1915. Joint missions between the USRCS and the U.S. Navy will also be considered. In that historical maritime period, the Revenue Marine and Revenue Cutter Service transitioned from wooden hull sailing (wind powered) cutters to auxiliary ships powered by sail and coal-fired steam vessels with iron hulls.

The wars the USRM/USRCS/USCG and U.S. Navy participated in included the Quasi-War with France (1800); the War of 1812–1815 against Britain; piracy and anti-slave ship wars; the Seminole Indian Wars; the War with Mexico (1846–1848); the Civil War (1861–1865); the Spanish-American War (1898); and the Great War (World War I, 1914–1918). The purpose and scope of this book will be to chronicle and analyze the evolution, missions, and leadership of the United States Coast Guard in that complex and significant period of maritime history. The chronological emphasis will be the 19th century age of sailing cutters and warships of the historic Coast Guard. The missions of the predecessor agencies of the USCG will be emphasized, with attention to joint U.S. Navy and USRCS missions, and the evolution of the USRCS from sailing cutters to steam (coal-fired)-powered cutters during the Civil War (1861–1865), the Spanish-American War (1898), and World War I

(1914–1918), the latter of which the United States entered belatedly (in 1917), approximately two years after the U.S. Life-Saving Service (USLSS) was merged with the USRCS in 1915 to form the United States Coast Guard.

This author's interest in the wind-powered sailing cutters was stimulated by the scholarly research and writing of the U.S. Coast Guard Atlantic Area historian William H. Thiesen, PhD, and his magnificent work on the U.S. Revenue Marine in the War of 1812–1815 against the United Kingdom in America's "Second War of Independence." Thiesen traced joint USRM-USN combat missions against the vaunted British Royal Navy in the Gulf of Mexico, the Atlantic Ocean, and along the Great Lakes and Chesapeake Bay. Dr. Thiesen's compelling narrative and cutter illustrations made me aware of the role of the Coast Guard in that extensive North American war, and the skill, courage and successes exhibited by the U.S. Revenue Marine against the premier 19th century naval and military power.

Readers of my previous books on U.S. Coast Guard history suggested that I write about the USCG, then called the U.S. Revenue Cutter Service, in the Civil War. The framework of this book on 19th-century naval warfare includes the war between the Federal Union and the states-rights oriented Confederacy and the political, geopolitical, maritime, military, naval, and technological complexities of that conflict.

The 19th and early 20th century span of this book begins with the 18th-century origins of the U.S. Revenue Marine under Alexander Hamilton's U.S. Treasury Department during the administration of President George Washington, the Quasi-War with France, the War of 1812 with the United Kingdom, and the subsequent 19th- and early 20th-century wars previously alluded to.

It is hoped that the reader, like the author, will come away with a renewed appreciation of the history and national defense responsibilities of the United States Coast Guard and its predecessor agencies and how that Service expanded its domestic and expanded responsibilities of maritime law enforcement, life-saving, aids to navigation, and national security and defense in an extensive historical process.

The multi-mission responsibilities of the sea service continued from World War I to the present, illuminating and solidifying its motto, "Semper Paratus" (Always Ready).

Introduction

The United States Revenue Cutter Service (USRCS) was initially referred to as the Revenue Marine. The Revenue Service is a predecessor agency of what would be become the United States Coast Guard in 1915 when the USRCS was combined with the U.S. Life-Saving Service.

On 4 August 1790, President George Washington signed an Act of Congress creating "A System of Cutters" of 10 small, armed sailing ships with a complement of 40 officers and men to collect customs duties on the tonnage of ships and goods and merchandise imported into the United States. The Revenue Marine was tasked with "securing the collection of revenue"[1] for the U.S. Treasury in this period of history before the creation of a federal taxation system.

This book will trace the history of the U.S. Revenue Cutter Service from its origins in 1790 to its amalgamation with the U.S. Life-Saving Service into the United States Coast Guard in 1915. The domestic peacetime missions of the USRCS will be examined, but the emphasis will be on the naval warfare and national defense history of the USRCS that prepared the way for the U.S. Coast Guard to perform its peacetime and wartime missions in domestic and overseas missions down to the present day.

The leaders, missions and ships, or "cutters," of the Revenue Marine, later called the Revenue Cutter Service, will be examined within the context of the several domestic and overseas combat missions the USRCS performed in partnership with the evolving United States Navy. The Revenue Marine was established under the U.S. Treasury Department and the leadership of the first Treasury secretary, Alexander Hamilton. "Cutters" were small wind-powered sailing vessels of one or two masts, armed with deck guns and sabers, cutlasses, and side arms. The cutters were named after British revenue vessels. Initially often less than 50 feet long, Coast Guard ships today are called "cutters" if they have berthing areas and are 65 feet or more in length.

The Revenue Marine, with its 1790 origin, predated the United States Navy authorized by Congress in 1795 and became combat active with its new fleet in 1798. The Navy and Marines were formed during the Revolutionary War (1775–1783) against Great Britain, but were disbanded by Congress commensurate with United States sovereignty and fears of standing armies and navies like Britain had. Then the U.S. Congress reestablished the Army and Navy branches in response to the maritime and military threats to the new nation posed by the naval fleets of the United Kingdom and France.

The U.S. Revenue Marine would assist the fledgling U.S. Navy in 1798 in the Quasi-War with France, as per the directive of the U.S. president, and establish the subsequent congressional and executive orders precedent that the U.S. Revenue Cutter Service and subsequently the U.S. Coast Guard support the U.S. Navy at home and overseas in time of war and international threats. The officers of the USRC *Massachusetts* and subsequent revenue cutters would enforce maritime law, especially tariff and customs regulations. Experienced seamen and officers were recruited from commercial vessels and the disbanded Revolutionary War Continental Navy.[2] President George Washington issued the first commission of command on the newly built cutters to Captain Hopley Yeaton. The earliest of the initial 10 revenue cutters were the USRC *Massachusetts*, *Diligence* and *Virginia*. These and other cutters and crews would board domestic and foreign vessels to enforce tariff laws and collect duties, battle pirates and smugglers, and enforce neutrality laws during the European Napoleonic Wars. In 1794 the USRC *Unicorn* captured a privateer that preyed on U.S. commercial vessels and was manned by sailors who supported France and the French Revolution.[3]

The domestic and foreign wars the U.S. Revenue Marine and U.S. Revenue Cutter Service were involved in during the 19th century were fought from 1798 to 1915. The Quasi-War with France (1798–1800, with the peace treaty being signed in 1801) occurred as the eight cutters of the USRM fleet were patrolling the Atlantic Coast and Gulf of Mexico and West Indies with the U.S. Navy. The cutters were classified in sailing terms defined by the number of masts and ship dimensions and guns and the kinds of weapons and armament. The cutters were brigs (2), schooners (5), and one sloop. The schooners and brigs each carried crew complements of 70 officers and men, with 14 guns. The sloop could carry up to 34 men and 10 guns.

The cutters fired upon and sometimes captured foreign military and commercial vessels. Cutters captured "prizes" alone (unaided) and also teamed up with USN vessels. The USRC *Pickering* captured 10 enemy vessels in West Indies waters, including one prize ship that carried 44 guns and 200 sailors, a much larger ship than the *Pickering*. The U.S. Navy waged war in the Mediterranean

Sea off the coast of North Africa against the Barbary pirates. U.S. Revenue Marine cutters did not cross the Atlantic Ocean to fight in the Barbary Wars (1801–1815), but several Revenue Marine officers volunteered to go with the USN into battle in the Mediterranean.[4] The Revenue Service was busy fighting pirates between 1800 and 1830 in the Gulf of Mexico and Florida waters. In August 1819 the revenue cutters *Alabama* and *Louisiana* battled pirate boats off Florida's southern coast in a contest that featured Revenue Marine crews boarding and battling the pirates in bloody hand-to-hand fighting. The maritime pirate-revenue cutter wars continued from the backwaters of Louisiana and into Gulf and Caribbean waters whenever the pirates ventured forth from their Latin American sanctuaries.[5] The other wars that kept the naval establishment occupied through the 19th century included the War of 1812–1815, the Seminole Indian Wars (1830s and 1840s), the war between the United States and Mexico (1846–1848), the Civil War (1861–1865), the Spanish-American War (1898), and the initial stages of the Great War, subsequently called World War I (1914–1918), which the United States entered in 1917.

The Revenue Marine articulated mission support, inspections, personnel training, transportation, and supply exchanges with the United States Lighthouse Service (USLHS) and the U.S. Life-Saving Service (USLSS). Those agencies are called predecessor agencies of the U.S. Coast Guard because the USLSS would be joined with the USRCS in 1915 to form the United States Coast Guard. The USLHS would be merged with the USCG in 1939, two years before the United States was officially involved in World War II.

In 1789, the first Federal Congress federalized the lighthouses that had initially been built and administered by the British in colonial America, some as early as 1716. The Congress would subsequently fund the construction, maintenance, and personnel of the lighthouses, including the construction, placement, and operation of what would be called aids to navigation. Navigation aids would include lighthouses, small boats and cutters called tenders, lightships, and buoys to mark channels, navigable waters, shoal (shallow) waters, and rocky portions of waterways.[6] Lighthouse keepers had the authority of a naval captain or commander over their crews and missions. Besides maintaining the warning lights to warn seafarers of shoal waters and to mark channels and harbor entrances, the lighthouse keepers—courageous females as well as males— organized and carried out life-saving missions.

The lighthouses were sturdy stone structures placed on the coast or on islands, in dangerous and isolated places. The crews were most active and their service most essential in times of darkness, high winds, heavy seas, and other inclement weather conditions. Where lighthouses could not be placed, government lightships and crews were anchored at strategic offshore locations.

The first anchored lightship was stationed on Chesapeake Bay on the eastern seaboard of the United States. Lightship duty was isolated, lonely, and dangerous.[7]

Maritime and Coast Guard historian John J. Galluzzo is a preeminent expert on the U.S. Life-Saving Service, its missions and personnel. Galluzzo, an author, lecturer, and writer, was the longtime editor of *Wreck and Rescue*, the journal of the U.S. Life-Saving Service Heritage Association (USLSSHA). Galluzzo has written extensively about the USLSS and its articulation with the U.S. Revenue Marine and U.S. Revenue Cutter Service and the merging of the USLSS and USRCS to form the U.S. Coast Guard in 1915. Galluzzo has described the Revenue Marine's response to privateers, smugglers and pirates and how French sailors used isolated U.S. shore locations as bases from which to attack and capture commercial vessels. Captain William Cooke, commander of the aptly named USRC *Diligence*, apprehended French privateers bringing gold ashore that had been confiscated from a Spanish ship. The revenue cutter crew returned the cargo to Spanish authorities. The author noted that the USRM acquired the responsibilities, of enforcing quarantines in ports besieged by malaria and flu epidemics, enforcing the slave smuggling embargo, interdicting illegal immigrants, and using lifesaving gear to assist mariners in distress on stormy seas.[8] The USRCS would become tasked with towing or destroying disabled ships and derelict vessels and performing coastal and river surveys.

The U.S. Life-Saving Service and its brave crews of "Serf Soldiers" traces its shore-based lifesaving traditions and lifeboats to the Boston Humane Society of the Commonwealth of Massachusetts, founded in 1786. In 1869, U.S. Treasury secretary George S. Boutwell and his staff merged the missions of the Revenue Cutter Service, the Steam Boat Inspection Service, and the Hospital Services to better protect and aid mariners and civilian passengers in distress and commercial and passenger vessels at sea. Secretary Boutwell made the fortuitous choice to appoint Sumner Increase Kimball, the chief clerk of the Treasury, to administer, coordinate, and support the missions of the personnel that saved more than 150,000 lives in maritime disasters between 1870 and 1915. Kimball assigned USRCS officers to inspect stations, train crews for beach patrols and sea rescues, and establish lifesaving stations along major river and coastal locations and on the Great Lakes. In 1878 the Treasury Department established the USLSS as a separate bureau and appointed Kimball its superintendent.[9]

The USRM teamed with the USN to battle the British in what some historians call the "Second War of Independence," the War of 1812, fought until the peace treaty was signed in 1815. The U.S. Navy battled the vaunted Royal Navy on the high seas and on the "Inland Seas" of the Great Lakes. The U.S.

Revenue Marine supported the U.S. Navy with shallow draft combat vessels and boats for coastal and interior maritime missions on lakes, bays and rivers. Treasury Secretary Albert Gallatin asked Congress to fund "small, fast sailing vessels to supplement the six available USN sailing frigates."[10]

Revenue cutters waged war with Royal Navy warships and British privateers in victory and defeat, proving that the crews and cutters were worthy adversaries of the Royal Navy and stalwart contributors to their U.S. Navy counterparts and colleagues in arms. Deck gun battles, bloody boardings, and on-deck combat were features of the maritime war, as were joint transportation, surveillance, and amphibious assaults conducted with the U.S. Navy, U.S. Marines, U.S. Army, and a variety of state militias. The USRC *Vigilant* and USRC *Eagle* were among the many cutters that earned glory in battle.[11]

The Revenue Marine engaged the Seminole Indians in the swamp and riverine regions of Florida in support of U.S. Army and U.S. Navy operations from 1836 to 1842. Indian warrior bands were battled on land, and civilian survivors were rescued. Troops, documents, and supplies were transported, rivers were blocked, and artillery and guns were distributed to soldiers and threatened settlements.[12]

The War with Mexico (1846–1848) found the USRC Service assisting the U.S. Army, Navy and Marines with coastal blockades, amphibious landings, and troop and supply transportation. The USRC *Jefferson Davis* accompanied U.S. Army infantry units in combat on the Pacific Coast. A decade after the Mexican War, the RC *Harriet Lane* joined an 18-ship U.S. Navy squadron. The shallow-draft, side-paddle, steam-driven cutter performed invaluable service to the U.S. Navy. Captain John Faunce, the commander of the *Harriet Lane,* received special acknowledgment in a report submitted by Commodore William Shubrick (USN) to the secretary of the Navy.[13]

The Civil War (1861–1865), as the Union called it, or the War Between the States, or War of Northern Aggression, as the states of the Confederate States of America (CSA) referred to it, was primarily a battle of land armies. But the naval forces of the USA and CSA played a significant role in the strategy, tactics, and outcome of the war. The naval battles were amphibious battles between the Confederate States Navy (CSN) and Confederate States Army on one side, and the U.S. Navy, U.S. Marines, and U.S. Revenue Cutter sailors on the other. In his classic book, *The United States Coast Guard, 1790–1915: A Definitive History* (1949), Captain Stephen H. Evans (USCG) stated that during the Civil War the Revenue Marine (or Revenue Service) was for the first time officially referred to as the "Revenue Cutter Service" in a law passed by Congress in 1863. At the time, cutters were still under the direct control of the collectors of customs in the ports at which specific cutters were stationed.[14] Evans traced the history of the USRM/USRCS from 1790 to 1915, describing cutters, missions, civilian and military leadership, and training. The initial cadet training occurred on revenue cutter vessels. Captain John Faunce (USRM/USRCS), of Civil War fame, was asked by Sumner Kimball, the civilian head of the U.S. Revenue Marine Bureau (1871–1878), to write the annual *Regulations* of the *U.S.R.M.* and the associated officer training and mission requirements. Captain Faunce responded with a masterful plan for an extensive civil, legal, and naval and liberal arts curriculum at the new U.S. Revenue School of Instruction (1877), the Revenue Cutter Academy. Captain Faunce outlined training procedures on the sailing revenue cutters *Dobbin* and *Chase*, and later on the steam-powered RC *Itasca*. The School of Instruction would transition from Arundel Cove, Maryland, to the former U.S. Army post of Fort Trumbull at New London, Connecticut, on the Thames, in 1910. Eventual groundbreaking would occur at the present U.S. Coast Guard Academy in New London, Connecticut (1931).[15]

Union and Confederate naval vessels consisted of a combination of ocean and river combat vessels and boats that were rowed, towed, or powered by coal-fired steam engines and sail (wind). Some vessels were wooden-hulled. Others were wood encased in iron and chain and were called ironclads. The steam-powered vessels were propelled either by a stern iron screw or side-wheel paddles. The sails were unfurled in open water to conserve on coal.

When the Civil War erupted in April 1861, and prior to that, during the secession of several states from the Union, a few Revenue Marine cutter commanders in Southern ports joined the Confederacy. U.S. cutters were assigned by the Treasury secretary to assist the U.S. Navy, give fire support to Union ground troops, assist in amphibious landings, and attack privateers and enemy ships, called commerce raiders, some of which were former Union cutters.

The USRC/USS *Harriet Lane* was in Charleston Harbor when the Confederate States Army fired on the U.S. military installation at Fort Sumter. That act officially started the Civil War, and the incident was therefore the immediate cause, as opposed to innumerable background causes, of the War Between the States (which the South also referred to as the War of Northern Aggression). The *Harriet Lane* therefore is credited with firing the first naval shot of the Civil War at a challenged ship. The USRC *Miami* would transport U.S. President Abraham Lincoln and several military and cabinet officials into the Confederate state of Virginia at Fort Monroe in 1862, to inspire Union forces to commence the Peninsular Campaign.[16]

Prior to World War I, the USRCS gained the naval combat experience that served the cutters and crews well in the Spanish American War (1898) and prepared them for the Great War. Eight revenue cutters armed with 43 guns supported the U.S. Navy blockade off the Cuban coast. The *Manning* engaged in combat in the Cuban theater, and the *McCulloch*, with a crew complement of 105 and armed with 6 deck guns, supported Admiral George Dewey (USN) in the Philippines. In the Cuban operation, the USRC *Hudson* went to the aid of the besieged Navy torpedo boat USS *Winslow*. The U.S. Navy assigned the duty of U.S. coastal surveillance to the U.S. Life-Saving Service via the numerous lifesaving stations.[17]

President Woodrow Wilson signed the legislation that merged the USLSS with the USRCS to form the U.S. Coast Guard for alleged reasons of mission, administrative, and economic efficiency. The stipulation that the USRCS could be transferred to temporary U.S. Navy control in national defense emergencies applied to the new Coast Guard. When World War I began in 1914, the U.S. Revenue Service cutters were assigned the mission of enforcing the neutrality laws of the United States. With the entry of the United States into the Great War in 1917, the cutters, shore stations, and personnel of the Coast Guard were placed under the control of the U.S. Navy. The transfer gave the USN the experience, support and personnel of the Coast Guard, which totaled 223 commissioned officers, 279 shore stations, 47 Coast Guard vessels, and 4,500 enlisted personnel.[18]

The violation of U.S. neutrality by German submarine attacks upon neutral vessels prompted President Woodrow Wilson to ask for, and Congress to issue, a declaration of war upon Germany and the Central Power allies on 6 April 1917.[19] It could hardly have been foreseen that the 10-cutter Revenue Marine created in 1790 under Treasury Secretary Alexander Hamilton to enforce the federal fund-raising provisions of the Revenue Act of 1789[20] would evolve into a military arm of the U.S. Navy in response to the belligerence of the European Central Powers.

The ability of the U.S. Revenue Cutter Service to merge with the U.S. Life-Saving Service to form the Coast Guard, and quickly respond to the Great War, was in large measure due to the experience and professionalism of the last captain-commandant of the USRCS and first commandant of the United States Coast Guard, Ellsworth P. Bertholf. Captain-Commandant Bertholf led the USRCS and USCG from 1911 to 1919. Advanced to flag rank as a commodore in World War I,[21] Bertholf was prominent in a historic line of distinguished civilian and military USRM-USRCS leaders, several of whose contributions will be traced in this book.

1

Cutters, Crews and Missions (1790–1915)

The Revenue Marine/Revenue Cutter ships of the 19th century were called "cutters," a nautical term used by Great Britain's Royal Customs Service. The U.S. Treasury initially referred to its revenue vessels as part of "A System of Cutters." The U.S. Coast Guard identifies cutters as any of their watercraft of 65 feet or more in length. The Revenue Marine cutters, and subsequent U.S. Revenue Cutter and U.S. Coast Guard missions, emphasized the enforcement of U.S. Customs laws and collecting federal duties. The USRCS and USCG missions gradually evolved into the other responsibilities of search and rescue, exploration, maritime law enforcement, contraband and immigration interdiction, ship inspection, icebreaking, scientific and federal official support, aids to navigation, environmental protection, and port and national security and defense missions. Coast Guard domestic, wartime, and overseas missions were often coordinated with the United States Navy.

A cutter was initially a sail, or wind-blown, wooden-hull vessel run by a seaman manning a steering tiller at the stern of the vessel. The first cutters had a full deck, one mast, a bowsprit and a gaff mainsail on a boom, with two jibs and a staysail.[1] The bowsprit is a spar extending from the bow or prow of the vessel, accommodating lines that secure the main or foresail. Some of the earliest cutters were authorized and funded by Congress. Various states and federal customs collectors operated other cutters. Some of the early cutters were simply port sailing or rowed barges and not seagoing vessels. The first ten (10) seagoing revenue cutters (USRCs, or RCs) authorized by Congress are listed below with launch date, building site and/or port location, and the shipmaster's name:

1. *Vigilant*: 1791, New York, Master Patrick Dennis;
2. *Active*: 1791, Baltimore, Maryland, and Chesapeake Bay, Master Simon Gross;

3. *General Green*: 1791, Philadelphia, Pennsylvania, Master James Montgomery;

4. *Massachusetts*: 1791, Newburyport, Massachusetts, Master John Foster Williams;

5. *Scammel*: 1791, Portsmouth, New Hampshire, Master Hopley Yeaton;

6. *Argus*: New London, Connecticut, 1791, Master Jonathan Maltbie;

7. *Virginia*: 1791, Norfolk, Virginia, Master Richard Taylor;

8. *Diligence*: 1792, Washington, North Carolina, Master William Cook;

9. *South Carolina*: 1792, Charleston, South Carolina, Master Robert Cochrane;

10. *Eagle*: 1793, Savannah, Georgia, Master John Howell.[2]

Howard I. Chapelle traced the history, missions, logistics, leadership, and characteristics of the revenue cutters of the U.S. Revenue Marine in his classic book, *The History of American Sailing Ships* (1935). This distinguished maritime author and expert on wind-powered sailing vessels marveled at the multiple mission responsibilities, knowledge, and leadership skills of the shipmasters and their crew members in challenging seas, changing weather conditions, and hazardous duties. He explained cutter construction and acquisition processes, and the scarcity of historical and shipbuilding records and reasons for it, and expressed his surprise at the relative lack of scholarly interest in the accomplished U.S. Revenue Marine/Revenue Cutter Service. The author described the Revenue Marine's contributions to seamanship, naval history, lifesaving, law enforcement, and national defense before the Service merged with the U.S. Life-Saving Service to form the United States Coast Guard in 1915. The author correctly noted that the USRM was the federal government's first navy before the U.S. Navy was created, and that it was manned, and mission ready by 1798, just in time for the U.S. Revenue Marine and U.S. Navy to combine forces against the French fleet in the Atlantic in the Quasi-War with France (1799–1801). Chapelle neatly synthesized the maritime missions and duties of the Revenue Cutter Service on all fronts: the enforcement of revenue laws, suppression of the slave trade and piracy, lifesaving, salvaging of ships and wrecks on the high seas, transporting government dispatches and officials, interdicting of smugglers and illegal immigrants, "and assisting the Navy in wartime, all with their many incidental adventures."[3]

Cutter shipbuilding records and blueprints are incomplete and difficult to access, given incidents of lost or burned records and because construction work in the age of sail was done in Navy shipyards or by private contractors. The USRM did not have its own construction agency or yard until the age of coal-powered steam cutters.[4] Cutter crews were generally skilled and competently led because cutters patrolled singly, as opposed to U.S. Navy flotillas and squadrons, so the commissioned officers had to be fit and self-reliant and to train their crews thoroughly. In single vessels, the officer had to be knowledgeable about maritime and revenue laws, seamanship, sailing in all types of weather and sea conditions, policing, elements of gunnery, the use of force, life-saving, salvage, the rights and dignity of free citizens, and public relations.

Secretary Alexander Hamilton advised his officers to be respectful of the citizenry and, during boarding procedures, aware of the importance of collecting revenue to fund the federal government and cognizant of the fact that smuggling was part of the American heritage during the colonial period of resistance to the oppressive taxation policies of British officials. Hamilton, the eclectic founder and leader of the Revenue Marine, studied and supervised shipbuilding operations and cutter characteristics to facilitate the missions of the Service. He planned and administered the service details, organization, appointments, personnel, sizes and classes of cutters, and the station and patrol regions each cutter was responsible for under the authority of the local port collector of customs.

Hamilton requested Congress to appropriate funds for the building and maintenance of ten cutters with initial keel lengths of 30 to 40 feet to carry a crew complement of two

U.S. Treasury Secretary Alexander Hamilton was the civilian head of the U.S. Revenue Marine at the inception of the Service in 1790 during the administration of President George Washington. The Revenue Cutter Service was formed as the "Fleet of Revenue Cutters" a few years prior to the U.S. Navy, with which it would partner in the Quasi-War with France and the War of 1812 and in subsequent wars and conflicts. The U.S. Revenue Marine, later called the U.S. Revenue Cutter Service, would evolve into the U.S. Coast Guard in 1915.

officers (a captain and a lieutenant) and six seamen. Armament would include small arms and six swivel deck guns. The local collector of customs was empowered to choose the design and builder of the cutter for that port until a captain was appointed to supervise and complete the construction project. Officer and crew courage, seamanship skills, and cutter seaworthiness were essential elements of mission success. Cutter speed and maneuverability were emphasized because the vessels had to operate in variable weather, shoal (shallow) waters, and also far out to sea. As missions and merchant traffic expanded, cutter sizes increased from approximately 50 to 110 feet in length, with a beam (width) averaging about 25 percent of the length of the vessel. The RC *Scammel,* for example, was launched in 1791 at Portsmouth, New Hampshire, as a two-masted schooner at a cost of $1,250. Her deck was 57 feet in length with a 15-foot beam, 6-foot cargo hold depth, and tonnage of 51.[5] The concept of the tons, or tonnage, of a vessel is a complex mathematical calculation of related concepts of the size and weight of a vessel, the weight of water displacement of a ship, and the internal volume or carrying capacity of a vessel.

One of the first ten cutters of the USRM was the RC *Massachusetts,* launched in 1791 for $1,440. The two-masted schooner was contracted to be 48 feet in length along the deck, with an 18-foot beam and a 7-foot cargo hold depth, at 63 tons. The completed cutter's dimensions slightly exceeded the contract dimensions. The two-masted RC *Vigilant* (1791) measured 48 feet along the deck, 15 feet in beam width, a 4'6" hold, and 33 tons. The RC *Active* (1791) was two-masted, able to carry 14 guns, and had a 40-foot length, 6'6" hold depth, 17-foot beam, at 47 tons. The RC *Virginia* (1791) matched the dimensions of the *Active.*

As missions expanded, cutter sizes and armament increased. Fortuitously, by the time the Quasi-War with France (1799–1801) commenced, the Revenue Marine had acquired ten newly constructed cutters and purchased three others. The new cutters measured 58 feet at the keel, 9-foot hold depths, 20-foot beams, and could carry from 10 to 14 deck guns.[6] The size of the cutters gradually increased in time because more of them were assigned as cruising cutters on seagoing patrols, as opposed to being confined to port anchorages and coastal patrols.

President Thomas Jefferson and his administration were reluctant to expand the naval and revenue marine fleets for economic, political, and ideological reasons, but they were gradually persuaded to do so because of the tense overseas relations with the North African Barbary pirates and the threats and attacks the pirates initiated upon American merchant vessels. Conflicts with Britain and France and the expansion of European navies persuaded President Jefferson to invest in naval ships and revenue cutters. It was fortunate that more

The USRC *Massachusetts* was among the first ten cutters constructed for the U.S. Revenue Marine. The U.S. Navy and U.S. Revenue Marine waged war on French naval and merchant vessels in American waters during the Quasi-War with France (1798–1800). U.S. naval forces in the war captured more than 20 French vessels. Revenue Marine cutters also captured slave ships and pirate ships in American and international waters in the early 19th century.

vessels were constructed between 1801 and 1812, because the United States would have to confront the British fleet in the War of 1812–1815, alternately referred to as the Second War of Independence.

Tracing cutter history is problematic because of limited and lost records and the practice of giving new cutters the names of previous ones. One such vessel was the Revenue Cutter *Massachusetts,* built in 1801 and revealed in an 1816 inventory to have the following characteristics: 58'6½" length; 7' depth; 62.5 tons; 10' draft; four lights on the main deck; assorted deck tackle, two boats, two rudders and oars for each; two 80-fathom (480 ft.) cables and two anchors; and a plethora of halyards, braces, ropes/lines, and sails. The two-masted cutter had a complex sail structure aloft and lift equipment. The RC *New Hampshire* (built in 1802–1803) was a sister ship with similar dimensions and characteristics of the popular *Massachusetts.* The schooner cutter *Louisiana* (1804) and RC *Jefferson* would distinguish themselves in action on various missions and in the War of 1812, as would Revenue Cutters *Surveyor, James*

Madison, Eagle, and *Vigilant.* These cutters would serve ably in combat with the U.S. Navy in the War of 1812, and some, like the *Alabama* and *Louisiana,* would serve gallantly, before and after that war, against encroaching European and Latin American privateers, pirate vessels, and slave ships.[7]

The revenue cutter stations and ports were primarily on the East Coast of the United States in the early part of the 19th century, but the RC *Louisiana* sailed out of the New Orleans station and carried out the usual Revenue Marine missions until 1830, including chasing down pirates and privateers from various Latin American nations. Other cutters joined the *Louisiana* in policing the Gulf of Mexico and areas around the insular entities of the Caribbean Sea. Cutters also sailed out of the Florida station of Key West.

In the 1820s and 1830s, a significant complement of U.S. Navy officers joined the USRM on leave from the USN, motivated by slow Navy promotions and a shortage of U.S. Navy ships to accommodate the commissioned officer ranks. The opportunity for junior USN officers to command their own vessels and carry out the challenging missions of the Revenue Marine stimulated the transfers. The migrating U.S. Navy officers soon chafed under the conflicting training requirements of the USRM, the need to work with and take orders from Revenue Marine officers, and to take instructions from civilians in the Treasury Department and politically appointed customs collectors in the cutter ports. To mitigate these conflicts, an order issued on 30 April 1832 required USN officers in the USRM to permanently choose between the two naval services. Some USN officers chose the USRM, and others either transferred back to the USN or were separated from the Revenue Marine.

In 1830, federal officials decided to build several larger seagoing (cruising) cutters that would accommodate an increased number of crew members and armament to facilitate partnership with the U.S. Navy as auxiliary warships and dispatch vessels but still serve effectively in port and coastal areas. The Revenue Marine consulted with Navy contractors and sought to have the new cutters constructed in U.S. Navy yards. Several cutters were built in the New York and Washington, D.C., Navy yards, but economies of scale and political and pragmatic considerations caused most of the new cutters to be built in private shipyards. Revenue Cutters *Morris, Gallatin,* and *Alexander Hamilton* were built at the New York Navy Yard. The statistical calculations originally determined for the RC *Morris* were for a 73-foot schooner, 20-foot beam, 7-foot hold depth, at 112 tons. Armament was to include six 6-pounder (weight of the shot) long guns, or six 12-pounder carronades.[8]

The U.S. Navy and U.S. Revenue Marine were the beneficiaries of armament evolution in the early 19th century that would be applied in the War of 1812, the Civil War, and subsequent wars. The warship weapons of the century

The USRC *Gallatin* was one of several revenue cutters built in the 1830s at New York City Navy Yard. The cutter was called upon to enforce federal tariff laws in Charleston Harbor in 1832 during the "Nullification" crisis and would, ironically, be captured by Confederate forces in the Civil War thirty years later.

included cannons, long guns, and pivot guns that would be used in the War of 1812. The Carron Company in Scotland contributed its name to the ship cannons of the period, called "carronades," that were first built for merchant ships (merchantmen) and the Royal Navy in 1776, during the American Revolutionary War (1775–1783).

Henry Foxall managed the Eagle Foundry in Philadelphia, Pennsylvania, on the Schuylkill River. Foxall is believed to have cast the first carronades for the U.S. Navy in 1799. Carronades were best used in close contact with enemy vessels because of range and accuracy paradigms. The guns were noted for their minimal weight and maximum power. The typical iron carronade weighed around 60 pounds, compared to the typical deck long gun of 150 to 200 pounds. The shorter carronade could be manned by just four sailors and fire a 42-pound shot from a wheeled carriage that would roll back about two feet, compared to the 10 or 12 foot recoil from the long gun. The carronade used less powder than the long gun. The shot moved at a slower speed and smashed out an irregular or jagged hole in the enemy's wooden hull. The resulting irregular shot entry hull hole caused more dagger-like splinters (shrapnel) to wound or kill enemy sailors and made a more difficult patch and repair task for carpenters on the target vessel. The weights of carronade shot varied from approximately 30 to 70 pounds. The shorter carronade range required the ship or cutter to fight at close range, a disadvantage if the enemy ship carried several long guns.

Nonetheless, the United States Navy and Revenue Marine ships and cutters that carried carronades achieved several significant victories at sea in the War of 1812 against the British Royal Navy and against the Confederate Navy in the Civil War. Civil War warship captains learned that it was best to carry both carronades and long guns in order to be ready for battle at various distances against different kinds of enemy vessels.[9] This was especially true in the Civil War, when both the USN and the CSN used formidable pivot guns and carried the long-range naval artillery cast-iron Dahlgren gun that fired shot several miles.

Contracts for more cutters were offered. Those cutters were built and launched in the 1830s by New York shipbuilders and the Washington, D.C., Navy Yard. Those cutters included the *Samuel D. Ingham, Louis McLane, Richard Rush, Oliver Wolcott*, and *Andrew Jackson* followed by the Revenue Cutters *Campbell, Dexter, Jefferson* (renamed the RC *Crawford* in 1839), *Roger B. Taney*, and *Washington*. The Revenue Cutters *Woodbury* and *Van Buren* were built in the late 1830s in Baltimore, Maryland. The Revenue Cutter *Erie* was built at Presque Isle, Pennsylvania, on Lake Erie.

The RC *Morris* would be in combat in the Mexican War (1846–1848). The RC *Gallatin* was active at Charleston Harbor, South Carolina, enforcing tariff and customs laws to counter that state's pre–Civil War Doctrine of Nullification (1832). The *Gallatin*, like several other revenue cutters, would be transferred to the Coast Survey in the 1840s. The RC *Hamilton* sailed out of Boston Harbor under the command of Captain Joshua Sturgis. The variety of Revenue Marine and revenue cutter missions is illustrated by the varied duties performed by the *Hamilton*: sailing to Nova Scotia to respond to the arrest of U.S. fishermen and chasing down and apprehending opium smugglers and vessels violating maritime laws. Captain T.C. Rudolph, an aggressive Revenue Marine professional who operated out of the port of Charleston, succeeded Captain Sturgis at the helm of the RC *Hamilton*. Rudolph clashed with the collector of customs who, like most of his political appointee peers, had little maritime knowledge or experience with cutters, cutter mission, and USRM personnel.

The Revenue Cutter *McLane* was known for its fine lines, masterful woodwork on the deck and gun carriages, and polished metalworks and deck guns.

Like many cutters that were damaged in heavy seas, tornadoes, and hurricanes, the *McLane* sank in a storm. The RC *Richard Rush* out of New York Harbor was heavily damaged in thick ice off New Haven, Connecticut, in January 1840. Even after the essential repairs, the vessel was considered unfit for sea duty and was assigned to the U.S. Lighthouse Service. The RC *Campbell* was damaged by a malady suffered by wooden-hull ships in Southern waters: dry rot. The cutter was sold in 1834, ended up as a slave ship, and was eventually taken into custody by a British warship.

The RC *Dexter* performed the usual variety of Revenue Marine missions until the U.S. Navy assigned the cutter to the protect coastal Florida settlements in the second Seminole War (1835–1842). The RC *Jackson* was stationed at the ports of New York and New Orleans, and then in Baltimore during the Civil War. The RC *Wolcott* reflected the diversity and geographic expanse of revenue cutter missions as reflected in that cutter's being assigned to several ports from the 1830s to 1851. The *Wolcott* sailed out of New Haven and New London, Connecticut; Wilmington, Delaware; Baltimore, Maryland; and

The USRC *Seminole* was built during the transitional age when naval vessels sported masts for wind-powered sails and stacks that spewed the smoke that came from coal and steam power. The cutter was named for the courageous Seminole Indians against whom naval and military forces waged relentless war in the swamps and along the coastline of Florida in the early to mid–19th century.

Mobile Bay, Alabama. Repaired in 1846 after being damaged in a Florida hurricane, the cutter was then assigned to carry official documents and dispatches on the Gulf of Mexico during the Mexican War of 1846–1848.

The RC *Jefferson*, under U.S. Navy jurisdiction, was in Florida waters during the second Seminole War and then wrecked on a point of land while on patrol out of New London, Connecticut. The RC *Madison* served in the Seminole Wars and was transferred to the Coast Survey in 1850. The RC *Taney*, whose namesake cutter would serve with distinction in World War II, prevented a revolutionary group in Florida from reaching Cuba (1851). The *Taney* later capsized in New York Harbor (1852). Following that incident, it was struck and heavily damaged by lightning off the Georgia coast and was finally sold in 1858. The RC *Washington* engaged in battles with American Indians along the Florida coast in the Seminole Wars and was later sold when hull inspections revealed extensive dry rot. In 1838, the RC *Woodbury* was attacked by a French warship ("man-of-war.") off the Gulf Coast of Mexico. During the Mexican War the *Woodbury* patrolled the Gulf of Mexico.

In the 1840s, several other cutters were built and performed significant civilian and naval service. The RC *Thomas Ewing* was a 170-ton sailing schooner that carried six ten-pound guns on deck and served in the Mexican War. After 1848 the cutter was stationed in San Francisco, California. The 150-ton RC *Walter Forward* displayed four nine-pound guns on her deck, served in the Mexican War, and patrolled for the Revenue Marine until 1865. Other revenue cutters that served with the *Walter Forward* in the Mexican War included the *Ewing, Morris, Van Buren,* and *Woodbury.*

Consideration of the use of steamships in the U.S. Navy and U.S. Revenue Marine can be traced back to 1837. Steam cutters were built in 1844 and 1845 and included the *Dallas, Jefferson* and *Walker.*[10] Sailing ship historian Howard I. Chapelle described the propulsion and the controversy of the first steam cutters: "They were fitted with the 'Hunter Wheel' or Ericson's screw propeller, but proved a most unsatisfactory craft. As a result of their trials, sailing and side-paddle-wheel cutters were built subsequently."[11] As Chapelle so aptly described it, the debut of the steam revenue cutters was disastrous and almost ruined the career of the Revenue Marine leaders who led the funding and construction efforts and put the cutters into service. More will be said about steam cutters and their more successful service in the Civil War.

Also serving the USRM in the Mexican War were the steam (coal)-powered cutters *Bibb, Legare, McLane, Polk,* and *Spencer.* USRM mission presence and challenges also occurred on the Great Lakes. The "Inland Seas" were unpredictable in the sharply divided seasons of the North Central states, and the humid continental climate of warm summers, cold and stormy autumns, and subarctic winters. The marginal presence of the USRM in the sparsely populated northern woodland waters posed a challenge to the Revenue Marine cutters and boats valiantly carrying out their assigned missions. The War of 1812–1815 was a major military threat in the region, as was the American Indian presence, most of the tribes being allied with the British and their Canadian subjects. The Revenue Cutter *Erie* served her namesake lake until 1849. In the late 1830s the *Erie* carried out Revenue Marine duties, including maintaining the neutrality of the United States during a period of rebellion in neighboring Canada, where rebels enjoyed the support of U.S. citizens anxious to aid the cause. The *Erie* crew carried out maritime and shoreside patrols to enforce U.S. neutrality and border security and was sold in 1849.

The cutters and crews described above, and those that will be considered in subsequent chapters, reflect the extraordinary range of missions and challenges faced by cutter officers and men in carrying out Revenue Marine and Treasury Department orders and responsibilities. From winter cruising in the Atlantic, rescuing mariners in distress, and salvaging damaged vessels (as ordered by Treasury Secretary Louis McLane in 1831 and Congress in 1837) to law enforcement and military and national defense missions, United States Revenue Marine cutters and crews operated in challenging and unpredictable circumstances. The skill and courage of the cutter crews are the stuff of which legends are made.

Cutter crews handled an increasing variety of missions as the cutter service increased capacities, personnel, vessels, and the geographic realms of their maritime domain. Naval service, winter cruises, lifesaving missions, furnishing equipment, food and water to isolated lighthouses, and supporting the U.S. Lighthouse Service by placing and maintaining channel buoys were among the assigned duties of the USRM. The naval support and what would later be termed the national defense mission required the upgrading of cutter armament after 1840 that included brass long guns (6-, 9-, 12- and 18-pounders) and 18-pounder iron carronades. The typical cutter of the period carried from four to six deck guns.[12]

In 1844 the Revenue Cutter Service was restructured to ameliorate previous administrative shortcomings, but the actions of inept customs collectors was problematic. Captain Alexander Fraser served as the chief of the USRM from 1843 to 1848. Fraser was one of the early "commandants" and his expertise and administrative skills, and the competence of his assistant, Lt. George Hayes (USRM), improved Revenue Cutter Service operations, cutter crew selection, and training. But issues with the initial steam/coal-power cutters and costs and repairs damaged Fraser's reputation in and outside the Service and prompted Congress to take procurement and innovation responsibilities from the Treasury

Department and place it under the control of appropriate congressional appropriations and oversight committees and subcommittees.

Several new cutters were built for Great Lakes, Atlantic Coast, and California ports, among them the USRC's *Campbell, Duane, Ingham, Harrison,* and *Joseph Lane.* The dimensions of the RC *Joseph Lane* measured 100 feet long, 23 feet in the beam, and a draft of 9-feet, 7.5 inches. The *Joseph Lane* and its sister ships featured improved crew quarters that included hammock supports and toilet provisions.[13] Several of these namesake vessels, like the *Campbell, Duane,* and *Ingham,* would compile distinguished combat records in their historic times, and their similarly named successors would honor their namesakes in World War II.

The California Gold Rush (1848–1855) increased the population of the Pacific Coast and required the Revenue Marine to purchase schooners and construct new cutters for the Western ports, among them the USRC *Argus.* In 1853 several new cutters were constructed in Massachusetts shipyards, including the *Dobbin* (a future Revenue Marine cadet training or "school ship" after 1876), *Cushing, McClelland,* and *Jefferson Davis,* a future hospital ship in the Pacific during the Civil War.[14] The *Jefferson Davis* was named for Jefferson Davis when he was the U.S. secretary of war (1853–1857) in the administration of President Franklin Pierce and not when he was president of the Confederate States of America during the Civil War.

In the antebellum period of the Civil War, and during the Civil War, the USRM acquired several additional cutters that were constructed, hired out, purchased, or borrowed. By the 1850s and 1860s Captain Alexander Fraser of the USRM had been vindicated as a visionary. The Revenue Marine, as of 1863 officially referred to by the federal government as the United States Revenue Cutter Service (USRCS), was using steam-powered, screw-propeller ships. In the past, the Revenue Marine had transferred obsolete vessels to the less restrictive U.S. Coast Survey, but during the period of cutter shortages in the Civil War the Coast Survey transferred some of its ships back to the USRCS.

Even as the USRCS transitioned to steamships, a few sailing cutters were constructed for the Service until 1878, when the last two, the USRC *Active* and the USRC cadet-training cutter (school ship) *Salmon P. Chase,* were completed. The *Chase* recorded a deck length a few inches over 106 feet, 25 feet in the beam, 11-foot draft, and more than 290 pounds of displacement. The *Chase* was under the initial command of Captain J.H. Merryman (USRM).[15]

Donald L. Canney, a noted historian of Revenue Marine and Coast Guard vessels, attributed the unique contributions of the USRM, and future U.S. Coast Guard, to their maritime history, safety, law enforcement, environmental protection, and oceanographic and national security duties. Those responsibilities facilitated the flexibility the agencies exhibited throughout their respective histories with limited budgets, assets and personnel. Canney described the Service missions as daily and diverse. Historical and geographic circumstances gave the U.S. Revenue Cutter Service and U.S. Coast Guard the international responsibilities of maritime safety, the International Ice Patrol, Bering Sea Patrol, maritime law enforcement, environmental protection, and scientific and oceanographic studies. Those missions stretched from the tropics and sub-tropics to the polar regions. Prohibition enforcement (1920–1933) expanded the missions and types of boats and cutters to supplement the enforcement arm of the U.S. Treasury Department and civilian Treasury agents ("G" Men). Those responsibilities required diverse and talented officers and enlisted personnel and the cutter and boat acquisition and design innovations needed to accommodate and enhance the seamanship skills required for the missions.

Researching cutter records is difficult, Canney contends, because of the loss of documents in fires and natural disasters and document dispersal to National Archives, Library of Congress, U.S. Coast Guard and U.S. Navy files. The construction of vessels in private and Navy and Coast Guard shipyards, and the helpful but extensive and diverse documents found in the 1935 Coast Guard *Record of Movements* files that chronicled the historical period 1790–1835 also illustrate the extensive and confusing milieu of sourcing, distribution, recovery and access. Canney attributed vessel identification confusion to cutter name changes and the practice of applying previous cutter names to subsequent cutters.[16] Given the challenges, Canney contributed a magnificent historical contribution with his research, writing, recording, and applying cutter history, diagrams, and primary and secondary sources in his 1995 book, *U.S. Coast Guard and Revenue Cutters, 1790–1935.*

Revenue cutters and Coast Guard cutters have traditionally (1790–1915) been named for the civilian U.S. Treasury secretaries, under whom the sea services served. The U.S. Revenue Cutter Service and U.S. Life-Saving Service were joined to form the U.S. Coast Guard in 1915. The Coast Guard would be under the U.S. Treasury Department until 1967, when the Service was transferred to the Department of Transportation (DOT), and subsequently to the new Department of Homeland Security (DHS) in 2003. Cutters have been named for a variety of entities, from floral life to Indian tribes to distinguished Coast Guard enlisted personnel and commissioned officers. After 2003 several of the planned new state-of-the-art national security cutters would be named after Coast Guard commandants and other USCG luminaries.

Between 1790 and 1890, according to Canney, the cutters named after Treasury secretaries were (by surname even though several cutter names included both given and surname): *Hamilton, Wolcott, Dexter, Gallatin, Campbell, Dallas,*

Crawford, Rush, Ingham, McLane, Duane, Taney, Woodbury, Ewing, Forward, Spencer, Bibb, Walker, Corwin, Guthrie, Cobb, Dix, Chase, Fessenden, McCulloch, Boutwell, Morrill, Sherman, Windom, Gresham, and *Manning.*[17] Famous cutters not named after Treasury secretaries, like the *Joseph Lane* and the *Harriet Lane,* will of course be covered in this book.

The revenue cutters built between 1790 and 1857 were predominantly in the "age of sail," or wind power, with a period of the introduction of steam vessels mentioned earlier that proved to be disastrous. Steam was not tried again until just before the Civil War. The first ten cutters were constructed in 1791 and 1792 and had one or two swivel deck guns, and small arms being issued to crew members. As cutters were built to exceed 80-foot lengths and 10-foot drafts and serve as auxiliary ships to the U.S. Navy, deck armament and ordnance increased in number and power. Cutters and crews gradually inspected and supplied the stations of the U.S. Lighthouse Service.

The Revenue Cutter Service abandoned the quest for steamships in the early 19th century, but as merchant ships increased in size and cutters increasingly served in seagoing (blue water) naval missions—and the sectional conflicts that would lead to the Civil War increased—the Revenue Cutter Service planners renewed their interest in steam-powered vessels. Attention was then focused on the construction of the highly regarded steam-powered side-cutter *Harriet Lane.*

The very first revenue cutter built for the Revenue Marine was the 60-foot *Massachusetts* (1791), with a complement of four officers and four enlisted personnel. The cutter was sold the following year and replaced by *Massachusetts* II. The RC *Argus,* launched in 1791, had a crew complement similar to the *Massachusetts,* with 4 swivel guns and a 50-foot deck. The *Argus* was home-ported at New London, Connecticut, the present site of the U.S. Coast Guard Academy. The RC *Pickering* was launched in 1798, captured five French vessels in the Quasi-War with France, and sailed with the U.S. Navy Squadron in the Atlantic and Caribbean. The *Pickering* was lost with all hands in a September 1802 gale.

Revenue Cutter *Eagle* III (1809) was purchased by the Boston Harbor customs collector and was under the command of Captain Frederick Lee. The cutter and her 25-man crew carried six 2- and 4-pounder guns. In October 1814, during the War of 1812, the *Eagle* courageously and effectively engaged the Royal Navy brig *Dispatch,* an 18-gun warship. Captain Lee put the heavily damaged *Eagle* on shore off the port of New York and continued the fight from a bluff. The *Eagle* was captured in an escape attempt the next day.

The RC *Louisiana* was built in 1819. Her construction plans were found in the National Archives in Washington, D.C. The *Louisiana* sailed the

Caribbean, where the cutter did slave ship and piracy interdictions. The 80-foot RC *Gallatin* was named for Albert Gallatin, President James Madison's Treasury secretary, who served as a U.S. diplomat at the signing of the Treaty of Ghent (Belgium) ending the War of 1812. The *Gallatin* served in both the United States Coast Survey and the U.S. Revenue Cutter Service. The cutter was captured during the Civil War by Confederate forces at Savannah, Georgia, and used as a privateer.

The RC *Jefferson* sailed for the U.S. Navy in 1836 in Florida waters during the Seminole Indian wars. The RC *Roger B. Taney* served as an inspection vessel for lighthouses and revenue cutters. The *Taney* was badly damaged by lightning strikes at the port of Savannah, Georgia, in 1858 and was sold. The 78-foot cutter *Hamilton* was constructed at the New York Navy Yard in 1830. Her crew complement ranged between 20 and 24. Cutters of her 112-tonnage displacement size typically carried four to six 6- to 9-pounder guns. The dangers in port and at sea in heavy weather inflicted significant damage on revenue cutters. The *Hamilton* was stationed in Boston Harbor under the command of Captain Josiah Sturgis and then at Charleston Harbor, where the cutter was lost in a December gale in 1853.[18]

Between 1858 and 1900 the USRCS expanded the size, crew complements, ocean-cruising capabilities, and armament and ordnance to match U.S. Navy capabilities in support missions. This expansion was in response to increasing domestic and overseas assignments in an era that put the armed forces of the United States into the Mexican War (1846–1848), Civil War (1861–1865), and Spanish American War (1898). In addition, with the U.S. purchase of Alaska from Russia in 1867 the USRCS commenced the patrolling and policing of Alaskan waters, the Bering Sea, and the Arctic polar region. Then the geographically stretched U.S. Revenue Cutter Service entered the Great War (1914–1918), during which time the USRCS and the USLSS merged to form the United States Coast Guard.

The USRC *Harriet Lane* was the best revenue cutter to exemplify the transition from sail to steam. The *Harriet Lane* would serve during the United States–Mexico War and the Civil War. Built and launched in 1857/1858, the 180-foot vessel was a 674-ton, two-masted brigantine with a 100-crew complement that eventually carried 10 deck guns of varying sizes that fired 24- to 30-pounders. Designed by the famous naval architect Samuel Pook and built by William Webb in New York, the *Harriet Lane* served in the U.S. Navy and performed gallantly in the Civil War before being captured by Confederate forces in 1863 and used as a privateer and blockade-runner. Captain John Faunce (USRM/USRCS), whose own distinguished career will be elaborated upon in a subsequent chapter, later recovered the cutter in Cuba.

Captain John Faunce, USRM, was the commander of the 180-ft. sail- and steam-powered paddle-wheel warship RC *Harriet Lane*, which fired the first naval shots outside Charleston Harbor, South Carolina, at the commencement of the Civil War on April 12, 1861.

Other cutters of the period include the RC *Johnson*, a side-wheel steam cutter similar to the *Harriet Lane* and designed for service on the Great Lakes. The *Johnson* was laid up during the severe winter months of that humid continental climate zone. The RC *Lincoln* served until 1874, patrolled Alaskan waters, and, like other cutters in the remote subarctic and Arctic realms, had official authority to establish and represent U.S. officials and the federal government in that isolated region.

Launched in 1861, the 350-ton, 153-foot RC *Delaware* was an iron-hulled steamer that carried 2 guns and a crew complement of 33. The *Delaware* served in the North Atlantic Blockading Squadron with the U.S. Navy in the Civil War. The RC *Gallatin*, launched in Buffalo, New York in 1871, was 133 feet in length, with a 40-crew complement and one deck gun. Assigned to the port of Boston, she sank in 1892. The RC *Grant* served from 1871 until it was sold in 1906. Armed with four guns, the 163-foot iron-hulled revenue cutter sported three masts, berthed a 45-crew complement and was stationed at the port of New York, then transferred to the Pacific realm, where she joined the Bering Sea Patrol in 1894. The *Grant* was the first U.S. revenue cutter to use wireless telegraphy communications, in 1903. The 115-foot USRC *Chase* was launched at Philadelphia, Pennsylvania, in 1878. The 32-crew complement manned 2 deck guns and served as a Revenue Cutter Service cadet training ship out of New Bedford, Massachusetts. The cutter sailed and steamed to Europe and then to the West Indies in the Caribbean Sea. In 1927 the *Chase* was transferred to the U.S. Public Health Service for duty as a hospital ship.[19]

The USRCS was active on the Great Lakes and coordinated inspection, transportation, and supplies with the stations of the U.S. Lighthouse Service and the U.S. Life-Saving Service. The Revenue Service, and later the USCG, worked with the USLHS and USLSS in a geographic evolution from east to west and from New York to Michigan, Wisconsin and Minnesota. When the 200th anniversary of the Coast Guard was commemorated in 1990, the Revenue Cutter Service and U.S. Coast Guard and their predecessor agencies had been at the far southwestern terminus in the Lake Superior twin ports of Duluth, Minnesota, and Superior, Wisconsin, for about 100 years. The U.S. Steam Boat Inspection Service was operating out of Duluth in the 1870s. The USLHS operated a light on southeastern Lake Superior on Whitefish Point in 1847 and at Duluth on Minnesota Point in 1857.

In 1873, the USLSS initiated the use of larger self-righting, pulling (rowing) and sailing lifeboats and surfboats modeled after the lifesaving craft of the British volunteer Royal National Lifeboat Institution (RNLI). Before the advent of gasoline-powered motor boats at the turn of the century, the modified sail-powered rowboats were helpful at the Great Lakes stations, where high surf and waves, heavy seas, and bad weather meant getting to endangered mariners and passengers far out on the lakes several miles from accessible roads. Surf men on the Great Lakes were seasonally challenged by winter snow and ice.

Among the cutters that sailed the Great Lakes was the RC *Fessenden*. This sleek, two-masted, iron-hull, paddle-wheel steamer was built in 1883 at Union Drydock Company in Buffalo, New York and replaced the wooden cutter of the same name. It patrolled the Great Lakes until 1903. With its 190-foot length, and 8-foot draft, this last paddle-wheel cutter on the Inland Seas was in service until it was sold in 1908.[20] The cutter sailed on Lakes Huron, Michigan, and Erie. After 1905, the *Fessenden* cruised out of Key West, Florida, into the Gulf of Mexico. In 1908 the cutter was sold, the last USRCS side-wheel steamer on the Great Lakes.[21]

The location of the USLSS and the Lifeboat Station at Duluth, Minnesota, dates back to 1866, the year after the Civil War ended. The annual reports for the Life-Saving Service for each of the years between 1866 and 1900 listed all of the USLSS Establishment Stations, their fiscal condition, equipment, personnel, mission statistics and operations. The busy lifesaving crews aided stranded merchant vessels and crews, including one that was carrying ore from Duluth to New York. Two cargo steamers collided about one mile from shore. All the crew members except one were saved. A damaged steamer in the Duluth Harbor was boarded, and the 20-man crew was saved. Dozens of rescues were made involving small boats, sailboats, and steamers in distress out on Lake Superior and in Duluth Harbor. In 1898 the crew of a sunken steamer was rescued, but three crewmen perished. In 1900, fourteen imperiled steamers were assisted, some with coal and ore cargoes worth thousands of dollars, the vessels worth multiples of that. The rescues that occurred between 1901 and 1915 in the final years of the USRC Service and the Life-Saving Service were significant.[22]

U.S. Lighthouse Service personnel saved ships and lives, but it would not be absorbed into the Coast Guard until 1939, when World War II erupted. The USLHS performed well from the time of its creation in colonial America in 1716 to its time in the United States on the nation's coasts, rivers, and Great Lakes.

Frank Albert Drew personified the skill and bravery of the personnel of the USLHS as Keeper of Green Island Station at Green Bay, Wisconsin, from 1909 to 1929. Green Bay is an arm of Lake Michigan. Drew had been the captain of a passenger and package boat but emulated his father by becoming head keeper at Green Island Light. Between 1912 and 1914 he rescued more than 30 people in storms, accidents, and fires on boats and ships. Captain Drew earned several lifesaving awards in his career. Marinette Marine Corporation in Wisconsin built the state of the art USCGC *Frank Drew* (WLM-557), which was commissioned in 1999. The 175-foot buoy tender has served the Atlantic coastal region out of Portsmouth, Virginia.[23]

The USRC *Bear* was built in Scotland, launched in 1874, purchased by the U.S. Navy in 1884, and transferred to the Revenue Cutter Service in 1885. The 198-foot, steam-powered and sail, 700-ton auxiliary cutter had a 50-crew complement and carried three 6-pounder guns. The cutter became a legend on the Bering Sea Patrol in Alaskan waters and in the Arctic on rescue, law enforcement, and federal government representation missions. The *Bear* served as a territorial courtroom, delivered Siberian deer to Alaskan Native Americans, and protected seal rookeries from hunting exploitation. The cutter and crews coped well in the Arctic environment, given the *Bear*'s iron-platted reinforcement within the wooden hull. The *Bear* and crew supported polar scientific missions and would go on to serve in World War II on the Greenland Patrol after the installation of diesel engines. In 1944 the *Bear* was decommissioned and in 1963 sank under tow off the Massachusetts coast.

The 148-foot iron-hulled twin-screw RC *Winona* (1890) had a 40-man crew complement and one deck gun and patrolled from North Carolina to the Gulf of Mexico. The 94-foot steel-hulled RC *Hudson*, commissioned in 1893, carried 11 crew members and two deck guns, was ported in New York City, and was assigned to the U.S. Navy in the Spanish-American War (1898). On combat patrol off the Cuban coast the *Hudson* towed the disabled and damaged Navy gunboat USS *Winslow* out of range of withering enemy gunfire.[24]

The Service experienced organizational and tactical modifications in 1886 and afterward. The USRCS now had its own engineers and naval architects. The new design and construction experts forged a partnership with the versatile Captain R.L. Shepard that led to the construction of the 15-knot cruising cutter *Windom* in 1896. The 170-foot cutter had more complex steam machinery, a steel hull, and advanced watertight compartmentalization as well as twin screws and a crew complement exceeding 40 men. Construction had begun at the Iowa Iron Works in Dubuque and was finished in Baltimore, Maryland. The *Windom* patrolled the mid–Atlantic coastal region until the Spanish-American War of 1898, provided naval support and blockade duty off Cuban shores during that war, then patrolled the Chesapeake Bay followed by patrol duty of the Gulf of Mexico out of Galveston, Texas, before the United States entered World War I in 1917. The *Windom* returned to the Gulf of Mexico after the Great War. The USRCS/USCG changed the RC *Windom* to the USCGC *Comanche* in 1915.

The gigantic 219-foot, three-masted RC *McCulloch* (1897) carried four 3-inch guns, a torpedo tube, and a crew complement up to 130, had a steel hull and planked wood, did 17 knots, had electric generators, and was assigned to the squadron of Admiral George Dewey (USN) that defeated the Spanish fleet at Manila Bay in the Philippines. The *McCulloch* patrolled out of San Francisco Bay in the Pacific before and after the Spanish-American War and served under the U.S. Navy in the Great War.

The 205-foot RC *Gresham* was constructed in 1896 at Globe Iron Works in Cleveland, Ohio. The *Gresham* had a 70-crew complement, six deck guns of various types, and a torpedo tube in the bow for antisubmarine warfare (ASW). The post–World War I presence of the torpedo tube when the *Gresham* was on the Great Lakes caused conflict with Canada due to treaties, signed between that nation and the United States soon after the War of 1812, that prohibited combat vessels on the Inland Seas. The Rush-Bagot Agreement, signed in 1817 and ratified by Congress in 1818, and the Webster-Ashburton Treaty of 1842 would be disputed and modified by British Canada, and later sovereign Canada, and the United States in subsequent periods of international conflict, war, smuggling, and terrorism.

The 188-foot Revenue Cutter *Seminole* (1900) had an 11-foot 8-inch draft and carried a complement of 8 officers and 59 enlisted men, and four 3-pounder deck guns. The *Seminole* patrolled from the southeast Atlantic Coast off the Carolinas, then transferred to the Great Lakes after World War I. The 207-foot RC *Nunivak* (1899) was a wooden-stern, paddle-wheel riverboat built in San Francisco and stationed on the Yukon River in Alaska. Hull damage caused her to be sold in 1902. The 188-foot wooden hulled RC *Thetis*, acquired by the USRCS in 1899, was, like the RC *Bear*, built in Scotland as a whaling ship, served on the Bering Sea Patrol, held federal court proceedings onboard, and shipped Siberian reindeer to Native American settlements in Alaska.

Launched in 1902 and commissioned in 1903, the unarmed 110-foot Great Lakes RC *Mackinac* patrolled the Sault Ste. Marie (Michigan) Locks and adjacent waterways before World War I. During the war, the *Mackinac* was transferred to the U.S. Navy for service in the Atlantic and returned the U.S. Coast Guard in 1919. Launched at the Newport News (Virginia) Shipyard in 1908, the 204-foot RC *Seneca* carried a 64-crew complement and four deck guns. The *Seneca* carried explosives with which to destroy derelict ships, cruised off the New England coast, and was assigned to the newly created International Ice Patrol to search for, map, and communicate iceberg locations to mariners. The International Ice Patrol was created by international agreement after the 1912 sinking of the British passenger liner RMS *Titanic*. During World War I the *Seneca* convoyed merchant ships in German submarine-infested waters between the United Kingdom and the Mediterranean Strait.

The diminutive 45-foot, 6-crew RC *Vigilant* (1910) was constructed in Bay City, Michigan. Powered by a gasoline engine, this harbor launch plied the waters of Sault Ste. Marie. The cutter and its mission illustrate the diversity of vessels the USRC needed to carry out its widespread missions and plethora of duties. Also exemplifying the multi-mission duties of the USRCS was the 152-foot RC *Snohomish*, launched in 1908 at Pusey & Jones Shipyard in Wilmington, Delaware. The *Snohomish* had a crew complement of seven officers and 46 enlisted personnel. The seagoing tug was outfitted with life- and property-saving equipment and carried surfboats and wireless communications. Like so many other revenue cutters before the completion of the Panama Canal, the *Snohomish* had to sail the east coast of South America and the treacherous and stormy waters around Cape Horn to get to U.S. Pacific Coast destinations.

The USRC *Miami* was launched by Newport News (Virginia) Shipbuilding Company in 1912, sailing the waters off the southeastern coast of the United States out of Key West, Florida, and carrying out winter cruising ice patrols in the North Atlantic. On 1 February 1916, the *Miami* was officially designated with a new name: USRC *Tampa*. With the entry of the United States into World War I, the *Tampa* was assigned to the U.S. Navy to accompany and guard ship convoys between Britain and the British colony of Gibraltar in southern Spain at the entrance to the Mediterranean Sea. The 190-foot vessel was armed with four 3-inch guns and successfully escorted eighteen convoys across U-boat–patrolled waters. On 26 September 1918 a torpedo fired by German submarine UB-91 struck and sank the *Tampa* with the loss of 127 military/naval and civilian passengers. There were no survivors. The sinking of the USRC *Tampa* was the second-largest loss of naval personnel in World War I.[25]

Before we leave this topic, a few words are merited about the agencies and personnel the USRM/USRCS articulated with. Coast Guard historian Dennis L. Noble chronicled a critical period in the Service: "The Kimball Years: 1871–1914."[26] Sumner Increase Kimball's appointment as chief of the Revenue Marine Bureau in the U.S. Treasury Department enhanced the foundation and advancement of the U.S. Life-Saving Service and the lifesaving stations fortuitously placed under his supervision in the bureau. He declared that he would accept the appointment by Treasury Secretary George S. Boutwell if the secretary backed Kimball's reforms against the expected political backlash of powerful members of Congress and special interest groups. Boutwell agreed to give Kimball the freedom to carry out the needed reforms. Kimball wisely chose Captain John Faunce (USRM), the distinguished Civil War commander, to inspect all the lifesaving stations and report on conditions, strategies for improvement, equipment, and personnel standards and training. In his 1871 annual report, and subsequent reports, Captain Faunce described dire conditions, of both stations and crews, and the reforms and standards necessary to make the USLSS functional.

The efforts and skills exhibited by Kimball, Faunce, and other supporters led to several technical improvements and innovations. Among them was the 1873 agreement with the U.S. Army Signal Service to establish telegraph lines between lifesaving

Sumner I. Kimball was the civilian head of the U.S. Revenue Marine Bureau and the U.S. Life-Saving Service and masterfully managed the Life-Saving Service until its merger with the USRCS to form the USCG in 1915.

stations on the eastern seaboard to facilitate more rapid communication and assistance procedures. Kimball persuaded Congress to fund USLSS operations, training and equipment, and to legislate ship safety, record keeping, and communications regimens to improve maritime safety and USLSS operations. After 1874 U.S. Life-Saving Service crews, stations, operations, technologies, equipment and surfboats were expanded on the Atlantic, Gulf, and Pacific Coast and the Great Lakes.

Kimball's meticulous record keeping made the Treasury Department, Congress, and commercial interests more aware of the dangers USLSS personnel faced and the successes achieved in saving mariners and vessels. On 18 June 1878 President Rutherford B. Hayes signed legislation creating the U.S. Life-Saving Service as an agency within the Treasury Department. The U.S. Senate confirmed Kimball as the superintendent of the USLSS. Inspectors of the life-saving stations would include U.S. Revenue Marine officers.[27]

George Putnam (third from right), standing with President Calvin Coolidge (center), was the pioneer commissioner of the U.S. Lighthouse Service and served in that capacity from 1910 to 1935. The USCG absorbed the USLHS in 1939, two years before the entry of the United States into World War II (Library of Congress).

Dennis Noble described and synthesized the mission articulation between the Life-Saving Service, the U.S. Lighthouse Service (Establishment), and the Revenue Cutter Service as well as the reciprocal support, inspection and supply missions offered by U.S. Navy, the U.S. Army, and the USRCS personnel to enhancing the lifesaving and aids to navigation performed by those services. George R. Putnam was appointed the first USLHS commissioner in 1910. Major David A. Lyle (U.S. Army) developed the famous Lyle gun, which threw the line used to connect disabled and wrecked vessels to shore and allow the transfer of equipment and people from ship to shore.

Noble deftly chronicled the political infighting between the Executive Branch, Congress, and various institutions, entities, agencies and interest groups that advocated for the merger of the various services, the union of the USRCS and the USN, and even the abolition of the U.S. Revenue Cutter Service.[28]

Howard I. Chapelle, the eminent historian of American sailing ships, concisely described the evolution and contributions of the United States Revenue Marine and U.S. Revenue Cutter Service: "The spirit and traditions of the present Coast Guard were founded on the traditions of the little revenue schooners of the days of sail. These rakish, topsail schooners, slashing through the heavy seas of a winter's gale, were the first to express the motto of the present Service, *Semper Paratus* (Always Ready)."[29]

2

The Quasi-War with France
(1798–1800)

The Quasi- [Undeclared] War with France commenced in April 1798 and ended with the signed Treaty of Mortefontaine in September 1800, which was ratified by the U.S. Congress in July 1801. The war was fought because of French maritime encroachments on United States sovereignty and the capture of American commercial vessels. The sea battles occurred off the Atlantic Coast and in the Caribbean Sea after Congress (in April 1798) authorized U.S. warships to seize armed French privateer vessels. By 1800 the U.S. Navy had seized several French privateers with the support of more than a dozen USRM frigates and sloops that were constructed at shipyards and also purchased merchant vessels. The Treaty of Mortefontaine, France (1800), required French naval and merchant vessels to stop capturing American ships. By the end of 1800 U.S. Navy secretary Benjamin Stoddert had secured more than 50 warships for the service, supported "by a score [20] of revenue cutters."[1]

Craig L. Symonds is the author of the *Historical Atlas of the U.S. Navy* and was distinguished professor of naval and Civil War history at the U.S. Naval Academy in Annapolis, Maryland. Professor Symonds also taught at the U.S. Naval War College in Newport, Rhode Island. His atlas is of inestimable value to students of naval history. Riveting descriptions of historical events, tactics, strategy, leadership, and technology enrich the magnificent maps in the *Historical Atlas*. William J. Clipson, Symond's Naval Academy colleague, used his cartographic skills to provide the atlas with the maps that cover the historical geography of the United States Navy. On his map of the Atlantic and Caribbean theaters of the Quasi-War with France, Clipson drew the West Indies islands of the Greater Antilles (Cuba, Haiti, Puerto Rico, and San Domingo) and the Lesser Antilles, which include numerous smaller islands. Symonds and Clipson identified the naval commanders, U.S. Navy warships, and Revenue cutters

34

involved in the insular battles. The map illustrates the stations of the eight revenue cutters in proximity to the islands, as well as U.S. Navy warships in proximity to where the cutters operated.[2] The British Royal Navy, U.S. Navy, and U.S. Revenue Marine ultimately persuaded France to sign the Mortefontaine agreement.

Robert Scheina wrote that eight cutters (two brigs, five schooners, one sloop) operated off the U.S. southern coast (Gulf) and in the West Indies. Cutter armament and crew complements ranged from 10 guns and 34 men to 14 guns and 70 crew members. Cutters captured eighteen prizes without Navy assistance between 1798 and 1800, and two others in joint USRM/USN operations. The USRC *Pickering* captured ten French vessels, including one vessel that was three times the size of the *Pickering* in guns and crew count. The Barbary Pirate Wars (1800–1815) occupied the U.S. Navy in the Mediterranean Sea and North Africa. The Revenue Marine did not send cutters overseas to

The 77-foot sailing brig USRC *Pickering*, which had fourteen 4-pounder deck guns and a crew complement of 70, fought in the Quasi-War with France. The cutter and crew captured five French vessels in the conflict.

fight the North Africans, but several officers of the Revenue Cutter Service joined the U.S. Navy and fought in the Barbary Wars.[3]

The United States got involved in the Undeclared War with France by agreeing with Britain to limit trade with revolutionary France and Napoleonic regimes while France and Britain waged war against each other on land and sea. Napoleon Bonaparte, the Emperor of France, needed the support of Europe's neutral nations. That diplomatic objective, and the impact of the surprising naval attacks upon French ships, motivated Napoleon to respect the sovereignty and neutrality of the United States and negotiate a settlement. President John Adams pursued the peace overture by sending American envoys to Europe to negotiate the agreement the United States eventually signed with France.[4] But before diplomatic ecumenism could be embraced, the war that forged the agreement had to be fought. The USRC *Pickering* served the United States and U.S. Navy well by capturing 10 French vessels in the West Indies, including the French privateer *l'Egypte Conquise* in a daylong battle against a ship that outmanned and outgunned the *Pickering*. Unfortunately, given the hazards of the open seas, the *Pickering* and crew disappeared in a storm in the fall of 1800.

Although the Royal Navy did not coordinate its missions with the U.S. Revenue Marine and U.S. Navy, the British shared supplies, ammunition, and recognition signals and allowed U.S. merchantmen to join escorted British convoys. American convoys reciprocated the favor. William H. Thiesen has done extensive research on the USRM in the Quasi-War with France and the War of 1812. Dr. Thiesen has served as the Atlantic Area Historian for the United States Coast Guard and has authored numerous maritime articles and a book about ship design and construction. The Revenue Marine and Coast Guard historian has also served on the executive staff of the North American Society for Oceanic History.[5] He has written extensively about the Undeclared War with France, the USRC *Pickering,* and the significant role the *Pickering,* its commanding officer, Captain Benjamin Hiller (USRM), and the USRM played in the war. The Quasi-War erupted only 14 years after the United States won its sovereignty from Britain in the Revolutionary War. Ironically, the French navy had played a significant role in helping America achieve its independence from the United Kingdom and in persuading United States policy makers to enlarge the U.S. Navy and the U.S. Revenue Marine.

Naval architect Josiah Fox made the plans for the cutter *Pickering,* the construction of which was funded by the U.S. Treasury Department. Built in Newburyport, Massachusetts, at the Merrill Shipyard in 1798, it was named after Secretary of State Timothy Pickering. The fast, two-masted cutter was a brig with a crew complement of 70 and was armed with 14 deck guns. Benjamin Hiller was one of the commanders of the *Pickering* even after it was transferred to the U.S. Navy during the war. Given the transfer, Hiller received the Navy rank of lieutenant in 1799. Between 1799 and 1800 the USRC/USS *Pickering* captured 15 French vessels, many of them holding more guns and larger crews than the cutter. The crew complement varied but had increased to 100 sailors and U.S. Marines.[6]

Some observers attributed Benjamin Hiller's earlier career appointment as a mate in the USRM to the fortuitous influence of his uncle, Major Joseph Hiller, a Revolutionary War acquaintance and postwar friend of George Washington. Major Hiller was appointed as the customs collector for the port of Salem, Massachusetts, in the 1790s.[7] Captain Hiller's successes during the Quasi-War with France earned praise in the New England news media for his disciplined crews, successful convoy missions, and feats of maritime warfare. U.S. Navy secretary Benjamin Stoddert commended the professionalism and contributions of "Commander Hiller." In a letter to Benjamin Hiller, the Navy secretary informed the cutter captain that President George Washington was also aware "of your merit."[8]

In the late summer of 1800 Captain Hiller left Delaware Bay to take the *Pickering* on another cruise to the Bahamas and the Caribbean. Unfortunately, the cutter foundered in the heavy seas and winds of a September hurricane, and the USS *Insurgent* and the RC *Scammel* struggled to survive it. The *Scammel* alone remained afloat after throwing anchors and deck guns overboard. The brave and hard-fighting crews of the *Pickering* and *Insurgent* disappeared beneath the waves, the *Pickering,* as William Thiesen wrote, "taking with it many documents, including bookkeeping records, letters, logbooks, and memoirs."[9]

The reports of a merchant vessel that sailed along that hurricane route a few days later reported floating vessel debris that matched the construction of the *Pickering*. Thiesen revealed that a Treasury Department fire in the "early 1800s, [and the] sacking of Washington by the British in the War of 1812, destroyed much of the archival material that was left to document the story of this historic cutter and crew," so the only sources of information available for historians are "contemporary accounts from newspapers, a few Treasury letters, and some naval records" from which to "trace *Pickering's* distinguished history."[10]

Among the archival documents that substantiate the record of the ships captured by the RC *Pickering* crews are those of Dudley Knox, the editor of *Naval Documents Related to the Quasi-War between the United States and France* (vols. 1–5, Washington: Gov't. Printing Office, 1935–1937). The dates of the captures are in 1799 and 1800, the number of vessels listed eighteen. Sixteen are named and two are unnamed. Commander Lt. Edward Preble (USN), commander of the *Pickering* in 1799, captured two of the ships. The remaining 16 vessels were captured when Benjamin Hiller commanded the *Pickering* in 1799 and 1800. The four captured enemy privateers carried from four to fourteen guns. Vessel types included brigs, schooners, an armed merchantman and a sloop. Two vessels were simply classified as "ships." The nationalities of the vessels were varied and mostly unlisted, except for an English "sloop" and a Danish "ship." Most of the vessels were probably French.[11]

William H. Thiesen included the RC *Eagle* in his 2008 *Sea History* article on Benjamin Hiller and the *Pickering*. Thiesen referenced a painting of the Revenue Cutter *Eagle* and the crew's capture of a French privateer in the Undeclared War with France. The *Eagle* saw naval combat in the Caribbean and off the Atlantic Coast. The *Eagle*, built in 1798, was commissioned as the USS *Eagle* when it was transferred to the U.S. Navy in that year. With a crew complement of seventy, the diminutive 58-foot cutter "captured or assisted in the capture of twenty-two (22) French vessels."[12]

Eight revenue cutters patrolled the southeast coast of the United States, the West Indies in the Caribbean Sea, and the Gulf of Mexico. The cutters ranged in crew complements from 35 to 70 men, with deck guns numbering from 10 to 14. Of the twenty-two prizes captured by the U.S. Navy and U.S. Revenue Marine in 1798 and 1799, Revenue cutters alone captured eighteen and cutters assisted the USN in capturing two others.

Before terminating this coverage of Quasi-War with France, a consideration of the diplomatic history of the war, its causes and its resolution may facilitate an understanding of the foreign and naval policies that drew the U.S. Navy and U.S. Revenue Marine into the war. In 1778 the Franco-American Alliance was signed between France and the United States during the Revolutionary War. The agreement affirmed a mutual defense pact and an understanding that the neutrality of the United States would be respected. But in the post–Revolutionary war administration of President George Washington (1789–1797) the United States declared its neutrality after the United Kingdom and France went to war in 1793. When the United States limited its trade with France, France retaliated. In 1796 and 1797 French privateers captured scores of U.S. merchant ships and approximately $200,000 in cargo and vessel value.

President John Adams (1797–1801) responded to the French capture of American merchantmen by funding the rapid completion of three U.S. Navy frigates: USS *United States*, USS *Constellation*, and USS *Constitution*. Harbor defense systems were planned and built, thousands of state militia (civilian soldiers) were armed and trained, and the federal government borrowed nearly three-quarters of a million dollars to fund the Quasi-War with France in 1798–1800. In exchange for a possible peace agreement, French officials contemptuously demanded the United States provide France with a $6 million loan and

$250,000 in bribe money. That insult to U.S. sovereignty resulted in the outright rejection of the French demands, and Congress gave 1,000 American privateers the legal authority (with official Letters of Marque and Reprisal) to seize or disperse French vessels on the high seas. Congress also levied a revenue tax on Americans, passed legislation to prohibit and prosecute antiwar civilian dissent, and established the U.S. Navy Department. The U.S. Revenue Marine was actually the first federal navy, established under the U.S. Treasury Department. The fledgling U.S. Navy and small U.S. Revenue Marine would combine forces in the so-called non-war. The U.S. slogan became, "Millions for defense, but not one cent for tribute." U.S. Navy warships and the smaller, faster revenue cutters would effectively challenge French maritime interlopers, just as the two sea services would once again the British during the War of 1812. That war was started when the Royal Navy also challenged American maritime rights, trade, and sovereignty.

By the end of 1798 the U.S. Navy and Revenue Marine had forced French vessels away from immediate U.S. coastal waters and into the Caribbean and Gulf regions. For political and diplomatic reasons President Adams refused to ask Congress for a declaration of war against France. The Royal Navy challenged the French navy overseas in the Mediterranean Sea, helping to keep elements of the formidable French navy out of American waters.

In 1799 Emperor Napoleon Bonaparte took political and military control of France and expressed a willingness to again recognize U.S. sovereignty and neutrality on the high seas. President John Adams responded by sending three U.S. diplomats to France to negotiate with Napoleon. French and American officials subsequently signed the Mortefontaine agreement in 1800.[13] Congress ratified the accord in 1801.

3

The War of 1812:
Background and Overview

The War of 1812 tested the sovereignty, national security, political leadership, and military and naval strength of the newly independent United States against the preeminent power of the United Kingdom. The United States had recently gained its independence from Britain in the Revolutionary War. In what came to be called the "Second War of Independence" the British were in turn testing that sovereignty and were determined to limit America's borders and autonomy

The political construct of federal leadership in the era of the War of 1812 included President Thomas Jefferson (1801–1809), his vice presidents (Aaron Burr and George Clinton), and Jefferson's secretary of state, James Madison. James Madison would serve as president from 1809 to 1817. Madison's vice presidents were George Clinton and Eldbridge Gerry, and his wartime secretary of state was future president James Monroe. Given that the U.S. Revenue Marine was under the Treasury Department, it is important to note that Jefferson's Treasury secretaries were Samuel Dexter and Albert Gallatin. President Madison's Treasury secretaries were Albert Gallatin, G.W. Campbell, A.J. Dallas, and W.H. Crawford. Gallatin would also serve as Madison's envoy to Ghent, Belgium, where the treaty that ended the War of 1812 with Britain would be signed. U.S. Revenue cutters will be named after several of these Treasury secretaries.

Some American historians trace the background stages of the War of 1812 to the Jefferson administration. As Britain and France renewed their war in 1803 Napoleon was induced to sell Louisiana to the United States that same year. The Louisiana Purchase included the vast Trans-Mississippi western territories, including land along the Gulf of Mexico. The Mississippi River and the city of New Orleans would be significant combat theaters in the War of

1812. Napoleon battled the British successfully in the European land war, but Admiral Nelson's Royal Navy defeated Napoleon on the highs seas, making Britain the dominant global sea power. Britain's naval power and maritime reach would lead to war between the United States and the United Kingdom. Britain would try to prohibit America's neutral ships from trading with France. The government of France also prohibited neutral trade with Britain and attacked U.S. merchantmen, as they had done in the Quasi-War with the United States. American merchantmen were thus stopped and boarded on the high seas by British warships and privateers, and cargoes were seized. Britain did not recognize the citizenship of American seamen of British heritage and impressed Americans into service in the Royal Navy and on merchant vessels to meet the needs of expanded maritime activities and to compensate for crew member desertion rates during the Anglo-French wars.

The assaults on the sovereignty of the United States, the economic dislocation caused by British attacks on American trade, and the impressment of American seamen inevitably forced President Jefferson to declare trade embargoes that hurt the American shipping industry and caused high unemployment. The U.S. Revenue Marine enforced the unpopular trade embargoes. On 1 June 1812, President Madison would seek a declaration of war. In mid–June, Congress consented, and declared war on the United Kingdom over that nation's seizures of American ships and sailors on the high seas and British violations of the neutral rights of the United States.[1]

The political and economic geography of the United States reflected regional differences. The New England states that furnished the ships and mariners for international trade generally opposed war with Great Britain. Senators and representatives from New England voted against the declaration of war. The national schism was such that some New Englanders appeared to favor that region's secession from the Union. Members of Congress and lawmakers along the Atlantic Coast from Virginia southward generally voted for the war against the United Kingdom and uttered sympathetic statements about the rights of seafarers and the sovereign right of free trade. Westerners and Southerners opposed British encroachments on shipping and trade and the impressment of seamen because their agricultural interests required exporting food crops, cotton, tobacco and hemp down the Mississippi River to the Gulf of Mexico and across the Atlantic unencumbered by the formidable British and French navies.

The War Hawks, as the pro-war politicians were called, passed bills to increase the size of the regular U.S. Army by 25,000, expand the U.S. Navy and Revenue Cutter Service, and use the war to invade and conquer British Canada. The War Hawks favored the acquisition of western lands as well and despised the British for still occupying U.S. territory and forts in violation of the Treaty of Paris (1783), which had ended the Revolutionary War. Americans also resented the British for agitating, arming and siding with certain American Indian tribes to stop American westward expansion. The British planned to form a Native American barrier and buffer zone between the United States and British Canada. Dynamic Native American leaders persuaded several Indian tribes to ally with Britain in the war and wage successful warfare against terrified Euro-Americans in the northern woodlands. Unfortunately for their cause, although the Indians did win several military battles, they would eventually lose their war against the American soldiers and settlers and forfeit their lands under treaties imposed upon them by numerically superior and better-armed white militia, the regular Army, and settlers.[2]

By the second year of the war, invasions of Canada by U.S. Army regulars and state militia had been turned back, and British Canada was "cleared of United States troops"[3] by patriotic, pro–British Canadians, many of whose ancestors had been pro–British Loyalists in the Revolutionary War. The Loyalists were forced to flee from that civil war in colonial America, and take up residence in British Canada. With Napoleon's abdication in 1814, the United Kingdom was able to transfer more naval and military assets to North America. The British war office shipped combat-hardened Royal Army and Royal Marine troops to U.S. and Canadian territory. Royal Navy transports and warships were sent to invade and blockade the United States along the Atlantic Coast, Great Lakes, Chesapeake Bay, and Lake Champlain in northern New York. The expansive body of water of Lake Champlain was strategically located close to the St. Lawrence River and Canadian boundary.

The British solidified its naval and military presence on the Great Lakes of Huron, Ontario, and Erie, and the St. Lawrence River watershed adjacent to Lake Champlain. The United Kingdom moved naval forces and troop transports from its Caribbean bases to attack New Orleans from the Gulf of Mexico, the Mississippi River watershed, and Lake Borgne. Lake Borgne is located at the approximate latitude of New Orleans. The vast geographic regions inexplicably did not moderate British optimism about its logistics and power, due to their arrogance as the superpower of that day and the lack of respect British war planners had for America's motivation and military and naval competence.

General Andrew Jackson, the controversial Indian fighter and future U.S. president, commanded American state militia regiments, regular U.S. Army troops, and an eclectic band of pirates, freed slaves, and Louisiana militia at New Orleans. Jackson made good use of the regional topography and swamps to build barriers against the expected British assault. The British force included more than 3,000 Royal Navy and Royal Marine personnel, dozens of warships, transports, and boats, and 11 transports that carried several hundred troops. The British forces were under the joint command of Major General Sir Edward Packenham and Admiral Alexander Cochrane. The British planned to defeat the American defenders and conquer the major port city of New Orleans and adjacent Louisiana territory.

General Jackson had 5,000 men, two sailing sloops that provided gun support at New Orleans, and seven gunboats with crews tasked with defending Lake Borgne. Between 23 December 1814 and 1 January 1815 American small arms and artillery stopped the British advance. On 8 January 1815 a frontal assault upon the New Orleans area defenses by more than 5,000 British military and naval personnel resulted in more than 2,000 of them being wounded, killed (including General Packenham), or missing. The Americans suffered a miniscule 71 casualties, including 13 killed and 58 wounded. The British withdrew the remaining forces to transport vessels for the voyage to West Indies bases and eventually back to the United Kingdom. Even though the Battle of New Orleans victory occurred after the Treaty of Ghent was signed on Christmas Eve 1814,[4] it would have significant geostrategic influence on the United States and the United Kingdom and their subsequent geopolitical relations and alliances.

By January 1815 U.S. and British officials and their citizenry were tired of the cost, bloodshed, and stalemate of the war. The Russian emperor offered to mediate peace talks between the Anglo-American nations, but Robert Stewart (Viscount Lord Castlereagh) offered to negotiate directly with the United States. President Madison accepted the proposal and immediately sent U.S. negotiators to the mutually agreed upon diplomatic site of Ghent.The American peace commissioners were Albert Gallatin, the recently resigned Treasury secretary and civilian head of the U.S. Revenue Marine, and John Quincy Adams, whose father had served as the second president of the United States. John Quincy Adams would be the sixth president of the nation from 1825 to 1829. The other envoys to Ghent were Henry Clay, James Bayard, and Jonathan Russell.

The British negotiators did not bring up several of the key issues that caused the United States to go to war: impressment of American seamen, arming Native Americans to establish an Indian buffer zone between the United States and Canada, and neutral rights on the high seas. The Treaty of Ghent conceded nothing and blamed neither nation. The concept of *status quo antebellum* prevailed in the treaty. Hostilities between the belligerents were terminated, the pre-war boundaries between the United States and Canada were affirmed, and any remaining and future boundary issues were to be settled in subsequent conferences.

American political leaders learned that the training of regular officers at the U.S. Military Academy at West Point, New York, was essential to the national defense, as opposed to relying on state militias and politically appointed or popularly elected officers. American leaders agreed that a strong and expanded U.S. Army and Navy and Revenue Marine were essential for the preservation of U.S. sovereignty and the protection of American commerce and seafaring rights on the high seas. Another byproduct of the War of 1812 was that national pride and unity were fostered in the United States and Canada, and the bonds of friendship and mutual alliance were enhanced between Britain and the United States.[5]

Historians Larry Schweikert and Michael Allen offered other insights about the regional politics of the war, General Andrew Jackson, the Battle of New Orleans, and the Ghent negotiations. The two historians contend that economic issues and military setbacks stimulated the antiwar movement in the New England states (Massachusetts, Vermont, Rhode Island, and Connecticut), and Maryland. The Federalist political party, in opposition to President Madison's Democrat-Republicans, convened at the Hartford (Connecticut) Convention in December of 1814 to discuss "Madison's War." The delegates examined political, foreign policy, and economic issues, and called for a separate peace between the New England states and the United Kingdom. Some of the delegates suggested that the New England states should secede from the union, a half a century before the Confederate States of America seceded and brought on the Civil War. With subsequent U.S. victories and a peace treaty with Britain, the Federalist political party fell into disrepute and disappeared with a quasi-treasonous image.[6] Some of the towns along the Chesapeake Bay had flirted with secession as well. Bad economic times and the threat of British invasion motivated some Americans to trade with the invaders. Some of that trade was in response to British threats to destroy towns, businesses and families.

The Treaty of Ghent did not specifically mention Native Americans, neutral rights, impressment of seamen, or freedom of the seas. The British and American envoys realized these issues would be resolved in the future by diplomatic negotiations. Fishing rights and British-Canadian and U.S. territorial boundaries were referred to future commissions and negotiations.

Schweikert and Allen dismiss the "David and Goliath" version of an outmanned and outgunned General Jackson at the ramparts of New Orleans, besieged by the overwhelming superiority of battle-seasoned professional Royal Navy sailors and British soldiers. The two historians contend that the skill and courage of General Jackson and his military colleagues were exemplary, but that his soldiers, sailors, and militias were combat veterans of the Indian wars supported by rough and ready French (Cajun and Creole), Spaniards, civilians, privateers, Choctaw Indians, and free black troops, accompanied by Jean Lafitte's notorious Caribbean pirates. General Jackson's forces were well protected by expertly built breastworks and surrounding open fields, swamps and forests, lakes, marshlands, and a canal, all of which the British had to navigate. Besides having to row through this unfriendly geography in small boats, British troops carried heavy (8-pound) cannon balls in knapsacks. A tipped boat meant drowned soldiers. The British also faced formidable, well-placed artillery and "sharp-shooting militiamen using Kentucky long rifles accurate at one hundred yards."[7] Exact casualty figures are disputed, but round-number averages are close to the following: American losses were 60 killed, 200 wounded, and 90 missing, for an approximate total of 350; British losses were 400 killed, 1500 wounded, and 550 missing, for an approximate total of 2,450.

Samuel Eliot Morison concluded that the War of 1812 was relatively inexpensive in monetary cost and human casualties and that international diplomats learned to respect each other, the United States became a respected world power, and most of the American wartime fleet was maintained after the termination of the war. Also, "within three months of the Treaty of Ghent [the U.S. Navy] found profitable employment in punishing [the] Barbary States for piracy." But despite those statements, Morison pessimistically concluded the following: "So ended a futile and unnecessary war which might have been prevented by a little more imagination of the one side, and a broader vision on the other."[8]

Schweikart and Allen disagree with Professor Morison. On the significance and results of the War of 1812, the historians concluded, "Jackson emerged a hero. Madison pardoned pirate Jean Laffite for his contributions, and the Federalists looked like fools for their untimely opposition. It was a bloody affair, but not, as many historians suggest, a useless one—'a needless encounter in a needless war,' the refrain goes. One conclusion was inescapable: the Americans were rapidly becoming the equals of any power in Europe."[9]

Walter R. Borneman wrote that the War of 1812 was "the war that forged a nation," a war fought between creaking sailing ships and inexperienced American armies often led by bumbling generals; American and British soldiers burning Canadian and American cities; a young nation of 18 states declaring war on the British Empire and chasing the British navy from Lake Erie "with a motley collection of hastily built ships," against a British Army and Royal Navy poised to retake Britain's former American colonies. The U.S. capital city of Washington and the President's House were put to flames, but a valiant defense in Baltimore drove the British off.

The British Royal Navy had blockaded major Atlantic and Chesapeake ports and tied up much of the U.S. Navy and U.S. Revenue Marine. But the two warring nations were eventually able to sign the Treaty of Ghent on Christmas Eve 1814, which preserved U.S. and Canadian boundaries and inexplicably avoided mentioning the major American grievances with Britain. The Battle of New Orleans was waged after the treaty was signed. The British were defeated. General Jackson would later be elected to "the U.S. presidency, and the United States would cast aside its cloak of colonial adolescence."[10]

Among the American diplomats who signed the Ghent Treaty was Henry Clay, a congressional War Hawk. Clay returned to the House of Representatives after Ghent and was reelected Speaker of the House.[11] Another delegate to Ghent was Albert Gallatin, the Swiss-born linguist who became a U.S. citizen, member of Congress, and founder of what would become New York University. Gallatin had been President Madison's Treasury secretary and, as such, the civilian head of the U.S. Revenue Marine. The revenue cutters of the USRM performed admirably alongside the U.S. Navy in carrying out its domestic and combat responsibilities. Gallatin assessed the significance of the war in his usual perceptive and articulate manner: "The war has been useful. The character of America stands high on the European continent, and higher than ever it did in Great Britain." Gallatin elaborated upon the influence of the war on the American character after a few more months of cogitation: "The war has renewed the national feelings the Revolution had given [that later] lessened. The people are more American; they feel and act more as a nation; and I hope that the permanency of the Union is thereby secured."[12]

The war stimulated a surge of nationalism and pride in British Canada as well. "Madison's War," as critics called it, revealed a land lust in the United States that mirrored if not mitigated the concern about maritime rights on the high seas. Frontiersmen favored expansion to the Mississippi River and north to the Great Lakes. Americans wanted to absorb Canadian lands and wage war on the British because Native American tribes blocked westward expansion, and the British were suspected of arming the Indians to block American expansion. Americans mistakenly thought Canadians desired independence from Britain and would flock to the American flag. Even Thomas Jefferson predicted that the conquest of Canada "will be a mere matter of marching."[13]

In 1814 four thousand Royal Marines marched into Washington and burned several buildings, including the abandoned White House and the Capitol building, which housed Congress. The ransacking and burning of Washington was payback for the devastation American troops delivered in the Canadian town of York, later named Toronto. The Canadians celebrated the expulsion of American troops, and the victory forged Canadian nationalism, pride, and unity. Military and naval victories and the Treaty of Ghent forged U.S. nationalism. Native Americans were the losers, because they were ravaged by the outcome, deserted by the British and outnumbered, overwhelmed and pushed off their lands by the migration of American settlers[14] with their dominating population, culture, technology, and weapons.

The War of 1812 is generally considered to have occurred primarily in the north central and eastern lands, lakes, and rivers, and off the Atlantic Coast of the United States. However, the Gulf Coast and Mississippi River regions in the southern and western sections of America were also significant theaters of war. In their classic book, *Westward Expansion: A History of the American Frontier* (Macmillan, 1982), professors Ray Allen Billington and Martin Ridge chronicled and analyzed the war and its impact on the southern and northern plains and the Mississippi watershed regions.[15] They describe the American easterners as patriotic nationalists whose agricultural and commercial pursuits galvanized their interests into a vivid support of the war and a willingness to take up arms on land and water to battle the invading British military and naval forces. After all, fur traders, farmers, plantation owners, and merchants were affected by Britain's support of marauding Indians. The Royal Navy posed a dire threat to the maritime transportation and agricultural productivity along the Mississippi River, tributary rivers, the Gulf of Mexico, and the Atlantic Ocean. Economic depression threatened the Ohio and Mississippi river basins and the West in general.

Western agrarians, fur trappers, and business proprietors needed the profits that came from the export of their production to European markets. Their bulky agricultural and commercial goods were shipped on the Mississippi River down to and out of the Gulf port of New Orleans. The variables of weather, floods, and pirates threatened river and Gulf transportation. The westerners were affected by the British embargo and the seizure of commerce-carrying merchant vessels. The frontiersmen blamed the British for arming and stirring up the Native Americans who posed a threat to westerners, although from the perspective of Native Americans the warriors sought to protect their people and their hunting, fishing, and croplands from Euro-American encroachment.

The westerners favored the U.S. invasion and conquest of British Canada to use as a bargaining chip to force the United Kingdom to allow free trade and stop supporting American Indian attacks. The landlubber westerners and their legislative representatives were not convinced that the few ships and limited numbers of personnel of the fledgling U.S. Navy and U.S. Revenue Marine could defeat the global reach and power of the Royal Navy, Marines and Army. The American conquest of North American Canada seemed to be the geostrategic answer.[16] Militant southerners favored using the war to acquire Spanish Florida for agricultural, commercial and demographical expansion and to subdue the runaway slaves, Indians and pirates who populated the region. Spain was too distracted from her own global conflicts to effectively police her Florida domain.

Citizens of the United States had already infiltrated Spanish Florida, and President Madison had declared with dubious logic and disputed facts that the region was actually U.S. territory as a result of the Louisiana Purchase in 1803. American military expeditions reached into Spanish Florida between 1810 and 1812. Western expansionists concluded that war with Britain and Spain would facilitate the acquisition and settlement of Florida.[17] The westerners' desire for the invasion and Northwest expansionists who thought boundary, fur trade, and American-Indian issues would be settled in America's favor by successful negotiations between the United States and an overstretched British government reinforced the idea of the conquest of Canada.[18]

From the perspective of Native American tribes, their attacks on frontier settlements and U.S. state militias and federal troops were essential to the preservation of their people and lands. As historians Billington and Ridge explained, "The Indian unrest that boiled through the back country by 1812 [was caused by the] land-grabbing treaties forced upon the natives by avaricious frontier officials and Indian agents who were instructed either to convert western Indians to agriculture or move them to unwanted lands beyond the Mississippi,"[19] into what historians and geographers would call the Trans-Mississippi West.

It was with gratitude and appreciation that the western frontier settlers greeted General Andrew Jackson's postwar Battle of New Orleans (1815) that thwarted the British attempt to conquer the major Gulf port city of New Orleans on the Mississippi River.

4

Naval Combat
in the War of 1812

Although suspicious of standing armies and naval power, President Thomas Jefferson decided to increase the number of revenue cutters to enforce the U.S. embargoes he supported and to meet the threat of British naval seizure of American ships and impressment of seamen. In 1809 Congress authorized the building or acquisition of 12 more cutters and more men. The cutters carried from six to ten light deck guns and fifteen to thirty crewmembers. The increase in cutter strength would prove fortuitous for U.S. Navy support when the War of 1812 commenced.

Among the significant cutters was the new USRC *Eagle* (launched in 1809) that served well in combat and gained fame by the gallant fight the crew put up against the British Royal Navy and the manner in which she was captured. Captain Frederick Lee (USRM) sailed out of New Haven, Connecticut, in response to Treasury Secretary Albert Gallatin's orders before and during the War of 1812 to enforce the commercial embargo laws, convoy merchantmen to New York Harbor, capture prizes, and "prevent the escape of vessels."[1] On October 10, 1814, the *Eagle* sailed out of New Haven to assist a merchant vessel reportedly captured by a British privateer sloop. In the dawn mist the *Eagle*, with its 6 guns and small arms (muskets), found itself in the proximity of the frigate HMS *Dispatch*, carrying 18 guns, including a 32-pound cannon. The *Eagle* escaped to the Long Island shore off New York City. The *Eagle* crew— plus the 40 volunteers, including militia, who had sailed with them—took the cutter guns and wrestled them up and onto a bluff and exchanged fire with the *Dispatch* from early morning until mid-afternoon. The *Dispatch* is said to have fired around 300 rounds against the cuttermen, who fired guns and muskets and even used the enemy shot that hit the ground around them. The cutter was badly damaged but saved from capture, as the enemy ship sailed away.

Unfortunately, the *Dispatch* returned the next day and captured the vulnerable cutter.[2]

The small size of the revenue cutters limited their ability to challenge well-armed and much larger Royal Navy men-of-war ships. But the cutters raised the morale of the newspaper-reading public who were all too familiar with British victories at sea. The cutters captured scores of enemy privateers and guarded American merchantmen from enemy capture and destruction. The first capture of a British merchant vessel occurred in April 1812, when Captain William Ham and the crew of the RC *Jefferson* seized a British brig and took it into Norfolk, Virginia. In July the RC *Surveyor* captured a British merchantman that had sailed out of the British insular colony of Jamaica. In August 1812 a South Carolina newspaper reported that the RC *Madison* (Captain George Brooks) captured the British brig *Shamrock*, which carried a crew of 16 and six guns.

But the often out-gunned cutters were sometimes lost at sea in battles and storms or captured in port by the large combat ships of the line of the British Royal Navy. The RC *Madison* was lost in 1812. The RC *Surveyor* (Capt. Samuel Travis) and his crew of 15 men were captured on a rainy night in 1813 while anchored in the York River. Fifty Royal Navy sailors with muffled oars rowed from HMS *Narcissus* (Lt. John Crerie, RN) to board the *Surveyor* and fight a bloody hand-to-hand deck battle using pistols (muskets) and other weapons. Casualties on both sides brought the conflict to an end. Lieutenant Crerie returned the captured surrender sword to Capt. Travis out of respect for the courageous fight put up by the cutter crew. In his June 13, 1813, letter to the cutter commander, the gallant British officer commended Captain Travis and his crew for a courageous defense of their vessel when greatly outnumbered, and concluded, "You have my most sincere wishes for immediate parole [from British custody], and speedy exchange of yourself and brave crew."[3]

The British privateer sloop *Dart* carried six 9-pound carronades and six deck swivel guns, more firepower than most revenue cutters, including the RC *Vigilant* (Captain John Cahoone). But that did not stop Cahoone from sailing out to meet the notorious enemy sloop that had captured 20 to 30 American vessels in the waters off New England and Long Island Sound. With 20 U.S. Navy volunteers on his cutter, Capt. Cahoone outmaneuvered the *Dart*, shot broadsides at the heavily armed vessel, and then boarded the enemy sloop and secured surrender. The exemplary service and naval support provided by the cutters and crews of the U.S. Revenue Marine led to the 1814 policy of putting battle injured Revenue Cutter personnel on U.S. Navy pensions.

Military action and vessel deterioration caused the loss from the ledgers of approximately 12 cutters, and the need after 1815 to construct or acquire

Commissioned in 1809, the 60-foot RC *Eagle*, with 6 cannons and a crew complement of 70, captured several British merchant ships. In 1814, after a fierce battle with HMS *Dispatch* at sea, the crew was forced to beach the cutter and fight gallantly from a Long Island bluff. The cutter was later captured.

new vessels. From this need came the gradual evolution of larger cutters (from 56 to 100-foot lengths and more), changes in the cut or shape of sails, and the replacement of the manned stern tiller with the wheel or helm forward on the deck. Armaments would vary but gradually included more functional seagoing cutters with 12- to 18-pound carronades, 9- to 12- to 18-pound long guns, brass deck guns, and amidship pivot-guns.[4] The increased capacities and crews of the cutters would prove advantageous when the Revenue Marine confronted pirate vessels and slave ships, Indians in the Seminole Wars, enemy soldiers in the Mexican War, and the Confederate military and maritime forces in the Civil War.

Military operations on land in the first year of the war were generally insignificant. The first American combat victories at sea came from the fledgling U.S. Navy and the U.S. Revenue Marine. The USRM and USN sailed on inland, coastal, and ocean waters to protect American commercial shipping and harass the combat frigates of the Royal Navy, which were generally larger and more heavily armed and sailed in greater numbers than U.S. naval forces. The USS *Constitution* (Capt. Isaac Hull, USN) put HMS *Guerriere* out of action in August 1812. The USS *United States* (Capt. Stephen Decatur, USN) took HMS *Macedonian* into custody in October. In December, the USS *Constitution*, under the command of Capt. Hull's successor, Captain William Bainbridge, USN, outgunned HMS *Java* in the South Atlantic Ocean east of Brazil. The following year (1813) saw diminished American naval action and limited success, as the British navy seized the upper hand with the disruption of American commerce, the blockade of port cities and coastal towns, and the seizure of the USS *Chesapeake*.

With the defeat of Emperor Napoleon in 1814, the British moved more army and navy personnel to the American combat theaters on the Chesapeake Bay, which provided a route from Virginia north to the Potomac River and Washington and Baltimore, Maryland. The British deployed more military and naval forces and equipment to the St. Lawrence River and Great Lakes maritime region between Canada and the United States and to Lake Champlain in northern New York State, adjacent to Canada. On 11 September 1814, Captain Thomas Macdonough (USN) used innovative ship maneuvers on Lake Champlain in the Battle of Plattsburg (New York) to defeat a British squadron on the expansive lake, a gateway to New York City and New England. Subsequently, the strong harbor defenses and skilled artillery tactics at Baltimore survived a British naval bombardment in that year.[5]

The Great Lakes were a major naval theater of operations in the War of 1812. The Inland Seas contain the fresh-water lakes Ontario, Ere, Huron, Michigan and Superior. Thousands of square miles of water, and each one consisting of hundreds of miles in length and width, the continental interior lakes were a formidable challenge to British and American boats and warships. Fierce storms, gales, fog, winter ice and snow, and gigantic waves that overturned top-heavy sailing warships posed significant natural threats. Ships could be, and were, marooned on sand bars and torn apart by uncharted rocks and shoal waters. Those waters, chilling depths, rapids, and rocky waterfalls posed navigation challenges and limited the areas open to combat. The lakes were so expansive shore batteries that could fire up to three miles (the outside limit of sovereign territorial waters) had limited functionality. Even Lake Champlain, part of the St. Lawrence River watershed, which includes the Great Lakes, was 14 miles wide at one point, and 125 miles in length. The lake system allowed access for troop and supply transportation for more than a thousand miles into the sparsely populated dense woodland interior of North America and the shorelines between the U.S. and Canada. The generally pro–British Indian allies terrorized land-based military personnel and the naval personnel who ventured on or too close to shore, as well as civilian farmers, townspeople, and laborers.

The woodlands provided timber for ship construction. The variety of trees met the diverse structural characteristics and needs of the hulls, masts, decks, and keels. The physical separation of the lakes, before the age of canals and locks, required that the naval forces of Canada, Britain, and the United States have three different building zones and locations for ships to be constructed and to patrol the main maritime theaters of war on Lake Erie, Lake Ontario, and Lake Champlain. Canadian/British naval and military infrastructure included the personnel and entities of the Royal Navy, Canadian Provincial Marine, customs and revenue vessels, and ports and forts. The Americans also populated the Great Lakes and St. Lawrence-Lake Champlain watershed with ports, forts, customs officials, regular troops and naval personnel, state militias, U.S. Navy ships and boats, and Revenue Marine boats and cutters.[6]

Mark Lardas, the distinguished historian of Great Lakes naval battles and ship construction in the War of 1812, has synthesized the formidable lexicon of sailing ship terminology that serves students of sailing ship history. Borrowing from the taxonomy, Lardas has defined the useful nautical terminology as follows:

- Brig: A two-masted, square-rigged ship with a fore mast and a main mast.
- Forecastle: Raised platform on fore (front) deck.
- Frigate: A sailing warship with full gun deck and other guns on forecastle and quarterdeck.

- Quarterdeck: Partial deck above main deck used for operations and navigation.
- Schooner: A ship with two or more masts rigged with fore and aft sails.
- Ship-of-the-line: A warship with at least two gun decks plus guns on the quarterdeck (toward the stern) and the fore deck (forecastle) intended to stand "in the line" of battle. Ships-of-the-line carried between 60 to 140 guns (the guns might pivot and be positioned amidships and along the side of the hull as broadside attack guns).
- Sloop: A small, single-masted ship with fore (front toward bow) and aft (toward the stern) sails.
- Sloop-of-war: A sailing warship with guns mounted on the gun deck, although guns could be mounted on the quarterdeck. Sloops could have one, two, or three masts.[7] The starboard side of the vessel is on the right facing the bow and portside is on the left.

Ship construction in the Great Lakes watershed was an enormous logistical undertaking. Woodlands had to be cleared to the beach and construction sites. Shipwrights, carpenters and laborers had to be drawn from faraway East Coast shipyards and coastal and local towns and cities. Laborers were drawn from rural regions closer to the fledgling shipyard sites that also had to be constructed. The Royal Navy, U.S. Navy and U.S. Revenue Marine would build warships or buy merchant or civilian-owned ships for conversion. During and after battles, captured vessels were then turned on enemy squadrons, flotillas, and enemy base camps, yards, and communities.

An example of the projects and sizes of the warships that were built in remarkably short periods of weeks and months is compelling. The USS *Lawrence* and USS *Niagara* were called brigs of war. These warships saw action in the strategic Battle of Lake Erie on 10 September 1813. The dimensions of two sister ships were 110-feet long; 9-foot holds and 10-foot drafts; 493-ton displacements; and crew complements of 135. The two warships were built at Erie, Pennsylvania, in the spring of 1813 and commissioned in August. They were armed with eighteen 32-pound carronades and two 12-pound long guns. The *Lawrence* and *Niagara* had stern davits from which a boat (gig) was suspended adjacent to the captain's skylight, which illuminated his quarters. The officers' wardroom was forward below deck and illuminated by a deck skylight. The foredeck featured a launch, and a smokestack extending up from the food galley located below deck. The crew berthing areas and hammocks were below deck, as were the sail and shot (ammunition) lockers. The shot lockers contained ordnance (powder, cannon balls, grape shot). Also below deck were the officers' cabin and bunks and structural and supportive equipment and supplies.[8]

Commodore Oliver Hazard Perry (USN) led the critical victory over the British Royal Navy on Lake Erie in the fall of 1813. The previous spring, then Master Commandant Oliver Hazard Perry assumed command of the naval base at Presque Isle on Lake Erie in Northwest Pennsylvania, about 25 miles west of the New York State boundary. Lake Erie stretched in a southwest-northeast direction from Michigan and Ohio and spanned northeast along the British Canada boundary and into New York State. Buffalo, New York, is located at the northeast boundary of Lake Erie, close to the 35-mile long Niagara River, which flows north to Ft. Niagara in northern New York on the south shore of Lake Ontario. About 20 miles north of Ft. Niagara across Lake Ontario was York (now Toronto), Canada. At the eastern end of Lake Ontario, north of Fort Oswego, New York, was the U.S. Naval Base at Sackett's Harbor. North of Sackett's Harbor was the British Naval Base at Kingston, Ontario. British and American naval commanders at those sites and at adjacent bases on the lakes periodically sailed out to exchange cannon fire and territory amidst maneuvering land forces. They generally broke off engagements after sporadic shows of force in order to avoid losing irreplaceable squadrons.

Commodore Perry succeeded Lt. Jesse Elliot at Presque Isle, but Elliot stayed on as second in command of the Lake Erie naval squadron. Perry appealed to the area military commander, Major General William Henry Harrison (U.S. Army), who loaned Master Commandant/Commodore Perry several U.S. Army personnel who had experience sailing on large vessels, and 100 U.S. Army and militia sharpshooters. Perry then sailed from Presque Isle on 6 August 1813, southwest to Put-In-Bay, to challenge the British at Detroit, Michigan and Malden, Ontario. Malden was located in proximity to Lake St. Clair and the Thames River in Canada. Captain Robert H. Barclay (RN) sailed his squadron out to meet Perry on 10 September 1813.

Commodore Perry quickly closed at a sharp angle with Barclay because Perry's flag vessel, the USS *Lawrence*, carried short-range carronades, and he wanted to avoid receiving broadsides from the British warship. The strategy collapsed when Perry failed to close the gap. The superior British gunfire caused Perry to lose most of his rigging aloft, and 80 percent of his personnel were killed or wounded. Perry ordered his sailors to row him in a boat between the active combat ships and over to the USS *Niagara*. Perry boarded the *Niagara* and relieved Lt. Elliot of command. Elliot had inexplicably stayed back during the battle and failed to close in to aid Perry on the *Lawrence*. Lieutenant Elliot was ordered to bring several small schooners into the fight against the British squadron. Perry managed to cross the bows of the two Royal Navy vessels he had battled from the battered *Lawrence* and won their surrender.

The outcome of the Battle of Lake Erie was significant. The British terminated their plans to invade Ohio. General Harrison went on the offensive against British land forces. Perry sent a scribbled note to Harrison stating, "We have met the enemy and they are ours: Two Ships, two Brigs, one Schooner & one Sloop." Perry then carried Harrison's troops across the lake to intercept British forces and achieve a military victory at the Battle of the Thames on 5 October 1813.[9] U.S. Naval Academy historian Craig L. Symonds summarized the significance of the Battle of Lake Erie: "Though Perry's victory did not lead to an American conquest of Canada, it did reverse the tide of battle on the western frontier and forestalled British claims to the Northwest Territory."[10]

A major maritime theater of war and the scene of multiple battles on land and sea was the expansive and strategic Chesapeake Bay. The British had used the vast waterway into the American interior during the Revolutionary War and would again do so in the War of 1812.

Standard geographic atlases and encyclopedias describe the southeast-northwest running Bay as a fresh water estuary extending from Maryland in the north to Virginia in its southern reaches and flowing into the Atlantic Ocean. More than one hundred tributary rivers and streams find their way into the Chesapeake in its course through a half-dozen states and Washington, D.C., on the Potomac River. The Potomac flows southeastward into Chesapeake Bay. The Chesapeake is approximately 200 miles long, with widths from about 3 miles to more than 20 miles. That extensive realm of maritime geography challenged the Royal Navy and its associated marine and army units. The vast body of water and its tributaries posed formidable logistical and tactical challenges to the U.S. Revenue Marine, U.S. Navy, and the soldiers of regular and militia units in their operations. With the many communities along the Bay, and the complex physical geography, it was difficult to sail and secure the Chesapeake watershed.

Tributaries of the Chesapeake include the York, Susquehanna, James and Patuxent rivers. The British Royal Navy, marine, and army units used the Patuxent River for their joint amphibious land and sea operations and for movement to what became the Battle of Bladensburg, Maryland. From Bladensburg, the British attacked the federal capital of Washington City in the fall of 1814. In September, the British went further up the Chesapeake and waged an unsuccessful attack on Ft. McHenry at Baltimore, Maryland.

Naval historian Craig Symonds and cartographer William J. Clipson aptly described and illustrated the War of 1812 theaters of war in their classic *Historical Atlas of the U.S. Navy* (1995). Symonds and Clipson covered the 19th century naval wars in scholarly detail and masterful narrative. Following Britain's defeat of Napoleon on the European continent in the spring of 1814, the United Kingdom was able to transfer more of its naval and military forces to the Chesapeake Bay, region where American forces were disorganized, dispersed, and weak. Commodore Joshua Barney (USN) moved his twelve barges and two gunboats south from Baltimore to meet the British fleet near the mouth of Chesapeake Bay.

Greatly outgunned and outmanned by Admiral Alexander Cochrane (RN) and his 20 warships and 4,000 troops, Commodore Barney redeployed to the north and up the Patuxent River to St. Leonard's Creek. At that site, U.S. Navy and Royal Navy barges engaged in a battle on 10 June 1814 that cost Barney one barge. The Royal Navy lost a sailing schooner that carried 28 guns. On 26 June, Barney's flotilla escaped from St. Leonard's Creek but was still threatened by the Royal Navy on the water and British troops on land. Because Barney's gunboats and barges were blocked by the British fleet, the commodore maneuvered north to Pig Point on the Patuxent River and abandoned his boats and barges.

In August 1814 Admiral Alexander Cochrane's 20 warships teamed up with Major General Robert Ross and 4,000 British regulars to face 7,000 American regular and militia troops, augmented by the 500 marines and sailors from the abandoned barge and gunboat flotilla of Commodore Barney. The confrontation erupted at Bladensburg, Maryland, on 24 August. The British rocket attack terrorized the retreating American militia, but Barney's naval personnel stood their ground and brought heroism and honor to the American defeat.[11] After Bladensburg, the British moved westward to Washington and set fire to several federal buildings, including the White House and the Capitol building, which housed the U.S. Senate and House of Representatives. The British fires were ignited in response to the American burning of York (Toronto), Canada, in 1813. The enemy attack on Washington caused Captain Thomas Tingey (USN) to incinerate the U.S. Navy Yard in Washington to prevent the British capture of the yard, its ships, and the weapons and ordnance. The conflagration destroyed the USS *Argus* and USS *Columbia*.

The British naval and military forces, victorious at Bladensburg and Washington, then returned to the Potomac River and sailed north to Baltimore, Maryland, to attack Ft. McHenry on the evening of 12 September 1814. The American attorney Francis Scott Key observed the British bombardment with mortar shells and rockets. The next morning, he gratefully observed that Fort McHenry's "flag was still there," despite the British "rocket's red glare," words that became familiar in what would become the lyrics of the United States national anthem. Despite their defeat at Fort McHenry, subsequent British victories in the Chesapeake Bay watershed in 1814 contributed to American demoralization and the horrible thought that the British might yet win the war[12] and possibly regain control of the former American colonies the United Kingdom had lost in the Revolutionary War and Treaty of Paris.

5

Revenue Cutters
in the War of 1812

As mentioned in the previous chapter, the USRC *Eagle*, commanded by Captain Frederick Lee, sailed out of New Haven, Connecticut, bristling with 6 guns and 40 crewmembers to confront the 18-gun HMS *Dispatch* on October 10, 1814. The outgunned and outmanned *Eagle* bravely challenged HMS *Dispatch* and suffered such heavy damage from 300 enemy rounds that the cutter had to ground on Long Island, off New York City. U.S. Coast Guard Atlantic Area historian William H. Thiesen described how the *Eagle* crew hauled ordnance up and onto a bluff and continued to battle the crew of the *Dispatch* with such bravery that the frigate sailed away to summon assistance. The next day, the *Dispatch* returned to the scene with another frigate and several armed barges. The two ships endured more firepower from the bluffs by Captain Lee and his intrepid crew but still managed to tie a line to the *Eagle* and tow it into captivity.[1] Captain Lee had captured several prizes before this clash with the HMS *Dispatch* and the Royal Navy. After his battle with the *Dispatch*, he said of his crew, "They did their duty as became American sailors."[2]

The historian William Thiesen chronicled the rest of Captain Frederick Lee's distinguished career and observed that the *Eagle* was one of six cutters lost in the War of 1812. One month after the RC *Eagle* was captured by the Royal Navy, a Boston newspaper reported that the cutter was convoyed to Halifax, Nova Scotia, in eastern Canada. In late December 1814, the customs collector at New Haven, Connecticut, paid the initial cost of $3,900 to build a new cutter *Eagle* that would patrol out of New Haven under the command of Captain Lee, who continued to serve in the U.S. Revenue Marine until 1829.[3]

The significance of the revenue cutter service in the War of 1812, and the impact of the war on the service was summarized by Thiesen: "Before the war, the revenue cutter fleet served primarily as the Treasury Department's maritime police branch (with the exception of combat action in the Quasi-War with France), enforcing U.S. trade laws and tariffs, and interdicting maritime smuggling. The *Eagle's* record in the War of 1812 helped establish the cutters new wartime missions of port and coastal security, reconnaissance, commerce protection, and shallow water combat operations."[4]

Paul H. Johnson was the curator of the United States Coast Guard Museum at the U.S. Coast Guard Academy in New London, Connecticut, when he wrote a 1977 article on Captain Frederick Lee in *The Bulletin*, the bimonthly magazine of the United States Coast Guard Academy Alumni Association.[5] For more than 10 years before his appointment as the Coast Guard Museum curator, Johnson was an academy librarian in Hamilton Hall. While he was there the historical murals painted in the 1930s by Aldis Browne fascinated Johnson. One of the mural panels featured an illustration of the Revenue Cutter *Eagle* under the command of Captain Frederick Lee in the October 1814 battle off Long Island and the Connecticut shore. The battle, described above and by Johnson in his *Bulletin* article, pitted the *Eagle* against a superior Royal Navy force.[6]

Paul Johnson applied his intellectual, research, and literary skills to a subsequent search for documents and other artifacts to illuminate Captain Lee's life. Johnson searched numerous regional archives and museums for more information. His own Long Island origins motivated his search for documents, maps, the location of the battle site, and his desire to "retrieve one of the buried abandoned cannons."[7] He searched through regional historical societies and Yale University for documents and charts of historic Long Island beaches and associated hydrographic information. A.O. Victor, the Yale curator of maps, revealed that at an auction he had purchased an engraved silver pitcher awarded to Captain Lee for his 1819 rescue efforts of the cargo and crew of the grounded merchant sailing ship *Betsey*. Victor had also acquired articles of Captain Lee's clothing and graciously donated those belongings to the U.S. Coast Guard Academy Museum in New London, Connecticut. The curator-librarian discovered information about Captain Lee's career as the captain of a merchant vessel and his appointment at the age of 43 into the Revenue Cutter Service in 1809. Lee was appointed by Collector of Customs Benjamin Lincoln to command the third revenue cutter named *Eagle* out of New Haven, Connecticut.

Lee was active at sea and ashore. The revenue cutter captain was a founder of the town of Madison, Connecticut, named for President James Madison. Lee led town meetings and represented Madison in the Connecticut State Assembly. Captain Lee founded an academy in the town that was named after him. Lee Academy still stood in the 1970s, when Paul H. Johnson published "In Search for Captain Frederick Lee." The academy building housed historical items. Johnson found Captain Lee's sword on display at the Whitfield Museum in Guilford, Connecticut, Lee's birthplace.[8] Even more astonishing was Johnson's discovery of a small, well-preserved portrait of Captain Lee painted by "the famous Polish volunteer in the American Revolution, artillery expert Thadeuz Kosciuszko who played a key role in the American victory at Saratoga (New York), perhaps the pivotal point," Johnson asserted, "in our struggle for independence."[9] The Polish artillery officer met and painted Captain Lee in 1797 while sailing as a passenger on Lee's commercial ship *Adriana* en route from Europe to the United States. Paul H. Johnson's curiosity and research skills, in his own words, "brought back from obscurity an early cutter officer" and illuminated "a glorious chapter in Coast Guard history."[10]

The 2 April 1813 edition of the *Charleston* described the in-port destruction of the Revenue schooner *Gallatin* (commanded by Captain John H. Silliman) anchored at the port of Charleston, South Carolina, on 1 April 1813. Captain Silliman was on shore after ordering that the cutter's small arms (muskets and pistols) be inspected and cleaned. At the time of the explosion, 35 crewmembers were onboard the *Gallatin*, with ten men on the quarterdeck and in the cabin, some of them inspecting and cleaning the weapons. Without warning, an explosion occurred that threw crew members through the cabin wall, hurling some into the harbor, and destroyed the quarter deck, tore up the main sail, demolished the stern of the cutter down to the water level, and sent fragments of the ship's spars aloft like shrapnel. Immediately, boats from shore and adjacent vessels came out to sea to assist in saving survivors as the fires on the cutter shifted to the rigging and mainsail. The wounded were placed in the rescue boats as the *Gallatin* sank stern first a few yards off the dock. Several of the Revenue Marine crew members were not found. Some of the wounded survivors "who were brought to shore," the *Charleston Courier* reported, "would have been happier if they had shared the fate [of their lost shipmates], as they cannot, in all human probability, survive the dreadful wounds they have received."[11]

The *Courier* article about the *Gallatin* explosion was written the day after the incident, when the exact cause of the conflagration was not yet certain. But perceptive reporters deduced that the fire originated in the magazine (ammunition) compartment."[12] It was discovered that the cabin door to the magazine had been left open while the crew inspected and cleaned the ship's stock of small arms. Acute maritime observers speculated that a spark or inadvertent gunshot struck the powder magazine and caused the explosion that sank the RC *Gallatin*. The newspaper called for a proper investigation to determine the cause of the accident, listed the names of the eight cutter-men that were missing, killed, or wounded, and revealed "an attempt will be made to raise the schooner."[13]

The Great Lakes maritime region was the scene of significant naval combat for U.S. Navy, Army, state militia, and Revenue Marine forces against British and Canadian naval and military units. The U.S. Treasury Department had established numerous customs collector ports and associated flotillas of Revenue cutters and boats to enforce federal customs and tariff laws, and the trade embargo. The numerous maritime clashes on the Great Lakes between the British Royal Navy and U.S. Revenue Marine and U.S. Navy vessels between June and November of 1812 illustrate the challenges. The Revenue cutters and boats were outmanned and outgunned by aggressive Royal Navy warships. On 18 June 1812 President Madison signed the congressional declaration of war against the United Kingdom that commenced the War of 1812. The congressional declaration stated, "The President of the United States is hereby authorized to use the whole land and naval forces of the United States … against the vessels, goods, and effects of the government of the United Kingdom of Great Britain and Ireland, and the subjects thereof."[14]

The Genesee (New York) District collector of customs, Caleb Hopkins, wrote to U.S. Treasury Secretary Albert Gallatin from near the site of present-day Rochester, New York, on Lake Ontario: "The shores being lined with Soldiers has induced me to dismiss all of my Deputies at this time, as not thinking them necessary, and shall wait your further directions [on] the subject."[15]

On 19 July 1812, the HMS *Royal George* and HMS *Prince Regent* captured the Sackett's Harbor (New York) revenue boat on Lake Ontario. The Royal Navy kept the prize cutter and paroled the captured Revenue Marine crew with orders to convey to U.S. authorities that the British would burn the American port unless a captured British brig was returned."[16] The British warship HMS *Prince Regent* captured the Ogdensburg, New York, revenue boat along the St. Lawrence River on 21 July 1812. In October and November British naval forces launched barges from HMS *Royal George* and captured the revenue boat stationed near Rochester, New York, on the Genesee River.[17] Two years later, from March through December of 1814, U.S. Customs collectors and revenue cutters and boats were again challenged. The *Plattsburgh (NY) Republican* reported that a revenue cutter on Lake Champlain in New York escaped a British attack. The customs collector at Oswego lost a revenue boat to Americans in the state militia who needed the vessel to defend an attack on the port city by British forces.

Peace talks between Britain and the United States began in August 1814. The Treaty of Ghent, which ended the war, was signed on Christmas Eve 1814. In February 1815 the warship HMS *Favorite* sailed across the Atlantic flying a flag of truce to bring a copy of the Treaty of Ghent to American officials in the port of New York City. On 16 February President Madison signed the Treaty of Ghent in Washington, thereby ending the War of 1812, before the postwar Battle of New Orleans victory in Louisiana by General Andrew Jackson had occurred.

On 25 February 1814, U.S. Treasury secretary Alexander J. Dallas informed all American collectors of customs by circular letter that cutter officers of the Revenue Marine "must be recommended for their vigilance, activity, skill and good conduct, ... smuggling must be prevented or punished," and customs and cutter officers and crews must immediately begin "restoring the light-houses, piers, buoys and beacons, within your district and jurisdiction, to the state in which they were before the war."[18] On March 3, 1815, the United States Congress repealed "the acts prohibiting the entrance of foreign vessels into the waters of the United States,"[19] and the Non-Intercourse and Non-Importation trade acts. On May 30, 1815, Secretary Dallas directed the customs collector of the port of New York City to begin building schooners to compensate for the revenue cutters and boats lost, captured, or destroyed on the Great Lakes-St. Lawrence-Chesapeake Bay watershed in the War of 1812.[20]

The "War of 1812 Revenue Cutter and Naval Operations" is an extensive document containing historical research from the U.S. Coast Guard History Program,[21] courtesy of the Coast Guard Historian's Office. The document reveals extraordinary details about the multi-mission responsibilities the U.S. Revenue Marine performed during the War of 1812. The Revenue Marine cutters not only carried out their traditional revenue collection, search and rescue, law enforcement, and exploration duties, but also participated in national defense missions with the United States Navy in the maritime domain of American inland and coastal waters. From April 4 to June 30, 1812, expanded cutter mission assignments challenged the Revenue Marine. Congress passed legislation that levied embargos on ships and vessels in American ports and harbors. The U.S. Revenue Service was lawfully required to enforce these unpopular laws. However, Congress did allow port departures of ships carrying federal government cargoes.

The RC *Thomas Jefferson*, under the command of Captain William Ham, sailed with Captain Stephen Decatur (USN) to survey lighthouses and establish signal towers for a military communications system that would use fire signals at night and flags during daylight hours. When Treasury Secretary Albert Gallatin notified customs collectors that war had commenced against the United Kingdom he ordered revenue cutters to notify all U.S. vessels in or off East Coast ports of the existing state of war. The RC *Massachusetts* (Capt. John Williams) was ordered by Secretary Gallatin to inform citizens of Maine that a state of war existed between the United States and Britain and the Revenue Marine would rigorously enforce the anti-smuggling laws that British and

American vessels were clearly violating. The South Carolina governor issued orders to customs and revenue officers, cutter crews and military officers of forts and stations in harbor areas to assist in the enforcement of trade and revenue laws and quarantine and health regulations. On June 24 and June 27 respectively, the RC *Thomas Jefferson* (Capt. William Ham) captured a British commercial brig carrying sugar from Guadalupe in the West Indies to Halifax, Nova Scotia. Between June 27 and June 29 Captain Daniel Elliot on the RC *Commodore Barry* seized a British merchant ship off the Maine coast, escorted three American schooners carrying British contraband to Portland, Maine, and seized another vessel off the coast of Maine for carrying British supplies.[22]

Continuing with Revenue cutter missions in the year 1812, the *Niles Register* of Baltimore, Maryland, reported that on the 4th of July the RC *Surveyor* (Captain Samuel Travis) captured a British brig and its cargo of coffee out of British Jamaica. The following day, the RC *James Madison* (Capt. George Brooks) out of Savanna, Georgia, assisted by navy gunboats, took custody of a captured British brig carrying turtles, British gold coins, and pineapples. On July 14, the RC *Eagle* out of New Haven, Connecticut, under the command of Captain Frederick Lee, reported to New York City naval authorities that the USS *Constitution* out-sailed a pursuing Royal Navy squadron. On August 1 the RC *Gallatin* (Captain Daniel McNeill) captured a British merchant vessel carrying contraband, including slaves, and brought the vessel into Charleston, South Carolina, for adjudication and prize money that would be divided between *Gallatin* crew members and officers.

Captain Daniel Elliott and his crew on the RC *Commodore Barry* confronted a four-ship Royal Navy squadron carrying more than one hundred deck guns. Captain Elliot beached the cutter at Little River, Maine, set up a battery composed of the cutter's guns, and fought five rowed barges of about 250 officers and men. After a two-hour battle, most of the outgunned Revenue Marine personnel made their escape into the adjacent forest. Two cuttermen (Charles Woodward and Daniel Marshall) were captured, becoming the first prisoners of war in the history of the Revenue Cutter Service and U.S. Coast Guard.

The Revenue Cutter Service performed search and rescue missions for endangered mariners at sea, and transported U.S. Army and state militia officers to operation sites. In their maritime missions, cutters were as vulnerable to storms and lightning strikes as merchant, military, and privateer vessels were. The Revenue Cutter *Louisiana* (Captain Angus Frazer) sank in a hurricane off the port of New Orleans in August 1812. Revenue crew members perished in the storm, but Captain Frazer survived to continue his war on Caribbean pirates and enemy privateers.

In the evening darkness of 22 August 1812, Captain George Brooks and the RC *James Madison* crew mistakenly challenged the 32-gun ship HMS *Barbados*, assuming, regrettably, it was just one of several merchant vessels in an enemy convoy. After a four-hour chase, the British warships overtook and captured the cutter in calm seas, using crew members and officers in rowed barges to board and seize the cutter. The captured revenue cutter personnel were placed on two of the Royal Navy warships. Captain Joseph Sawyer, commander of the RC *General Greene*, and his crew, supported by sailors from a navy gunboat, boarded the American privateer *Superior* to find that the vessel was carrying British contraband goods. A quarrel ensued between the New Castle, Delaware, customs collector and the regional navy commander. A subsequent federal court decision (June 28, 1815) awarded legal possession of the *Superior* to the customs collector because the vessel was seized in U.S. waters in violation of trade embargoes and was therefore under the legal custody of the customs collector and the revenue marine officers. In September of 1812, in response to the active patrols of the RC *General Greene*, the Treasury Department approved an increase in the cutter's crew complement to 24 men.

The revenue cutters maintained their active patrol schedules during the final months of October, November, and December of 1812, as indicated by the successes and failures of the following incidents. In October, a revenue cutter small boat was captured by men on barges from the 10-gun HMS *Royal George* near Rochester, New York, at the mouth of the Genesee River. In one cutter capture by the British, nine enlisted Revenue Marine prisoners were sent to Halifax, Nova Scotia, four to Boston, and the rest, including several black prisoners, to a prison in England. In October 1812 cutter Master Angus Frazer completed his service with a U.S. Army reconnaissance team that had chased smugglers and officially surveyed Louisiana bayous and insular geography. Earlier that month, Captain Frazer, the former commander of the RC *Louisiana*, had arrested armed smugglers and captured their boat. The smugglers escaped and returned the following day to recapture their boat and take Frazer and his crew into custody, but the mariners made their way free the next day.

A newspaper report published on 5 November 1812 revealed that a revenue cutter in the Northwest Territory of Michigan transported U.S. Army general William Hull on several missions before the former governor was defeated by the British in battle. The remainder of November included an incident where Captain Joseph Sawyer and his RC *General Greene* crew used axes to cut through the wooden hull of a commercial brig and rescue a trapped crew in a storm off the port of Philadelphia. On November 24 a New York newspaper reported that the captured officers from the RC *James Madison* were released in New York City. On 29 November 1812 the RC *Diligence*, under the command of Joseph Burch, rescued the survivors of a merchant ship that capsized in a storm off the treacherous waters of Cape Fear in North Carolina. The *Diligence* crew helped bury the dead and salvaged some of the brig's cargo.[23]

The second year of the War of 1812 offered no respite for the Revenue Marine sailors and cutters. On 31 January 1813 the crew of the RC *General Greene* (Captain Joseph Sawyer) saved the prize ship *Lady Johnson* from destruction and shore-side damage as it drifted in thick Delaware Bay pack ice. The adjudication process in the Delaware district court concluded that the captain and cutter crew "exposed themselves to the rigors and severities of a most inclement season, succeeded in removing the said ship to a safe place ... where she now lies [with] part of her cargo on board in perfect safety."[24]

The British fleet under Rear Admiral George Cockburn (RN) escalated the threat to U.S. naval and military forces, and maritime communities when, on 4 February 1813, the entrance to Chesapeake Bay was put under blockade. Secretary Albert Gallatin responded by ordering the Norfolk, Virginia, customs collector and the Revenue Marine to extinguish navigation lights and secure and remove lamps, oil, and other lighthouse mechanisms on Chesapeake Bay to prevent the use of the navigation aids by the British. On 20 March 1813 the Wilmington, Delaware, customs collector ordered Captain Sawyer and the crew of the RC *General Greene* to observe and report Royal Navy locations and activities in the blockade of Delaware and Chesapeake Bay blockades. Ten days later Admiral John Warren (Royal Navy) extended the British naval blockade to include the American port cities of New Orleans, Louisiana; Savannah, Georgia; Charleston, South Carolina; and New York City.

On 11 April 1813, the RC *Thomas Jefferson* (Captain William Ham), with assistance from local militia, captured three Royal Navy barges and 60 commissioned and enlisted crew members. The British barges had been pursued up the James River of Virginia and threatened with the broadside cannon fire of the RC *Thomas Jefferson*. The *Alexandria Gazette* reported that Captain Ham ordered the British sailors ashore to be placed in the custody of militia riflemen. During the month of May, British naval forces set fire to three Maryland towns.

The New Haven (Connecticut)-based RC *Eagle* (Captain Frederick Lee) served as the transfer vessel for a prisoner exchange between British and American forces. More than 30 Americans were released from British custody. On 26 May a New York newspaper reported that the RC *Active* (Captain Caleb Brewster) did reconnaissance on three British warships at sea and forwarded that intelligence by small boat to Commodore Stephen Decatur (U.S. Navy), and his Long Island Sound flotilla. On 28 May 1813, prisoner of war John Barber, from the RC *James Madison*, became the first member of the Revenue Marine to die as a captive on a British hospital ship (HMS *Le Pregase*) off the port of Chatham in the United Kingdom. On 8 June 1813 the RC *Active* evaded the Royal Navy's blockade at New London, Connecticut, to team up with Commodore Decatur's U.S. Navy flotilla on the Thames River. However, two days later, a Royal Navy squadron was able to blockade the U.S. Navy flotilla, including the USS *United States*, USS *Hornet*, and RC *Active*.

On 25 June 1813 British troops landed at Hampton, Virginia, attacked and damaged the city, and used the captured RC *Surveyor* to support the operation. On 12 July the RC *Mercury* (Captain David Wallace) evaded an attack by 15 armed barges and 1,000 British Navy personnel. The *Mercury* carried important customs documents and bonds to New Bern, North Carolina, and prepared the town for the coming attack. The local newspaper praised the men of the *Mercury* for their timely warning.

On 15 September 1813 an officer from the British warship HMS *Pears* informed the deputy customs collector of Ocracoke, North Carolina, that the port and several to the south, and all ports south of Boston would henceforth be under British naval blockade.

On 7 August 1813 Captain Samuel Travis, the former commander of the captured RC *Surveyor,* returned to Virginia after being paroled as a prisoner of war by the British. On 24 August 1813 acting Treasury Secretary William Jones authorized the New Orleans, Louisiana, customs collector and U.S. Revenue Cutter officers to work with U.S. Navy officers in the apprehension of smugglers, slave traders, and pirates in the Lake Barataria region. Another example of U.S. Revenue Marine and U.S. Navy cooperation occurred on 3 November 1813, when the RC *New Hampshire* (Captain Nathaniel Kennard) out of Portsmouth joined Commodore Isaac Hull's flotilla of two U.S. Navy gunboats and the warships USS *Enterprise* and USS *Rattlesnake* to pursue Royal Navy brigs. In December of 1813, in an attempt to stop the flow of illegal trade between American smugglers and British naval and military forces, the U.S. Congress passed an act to prohibit American merchant ships from leaving the ports and harbors of the United States.[25]

The year 1814 offered continued challenges for the Revenue Cutter Service in the war. The Baltimore press reported in January that Captain William Ham, on the RC *Thomas Jefferson*, sent an American schooner ashore and into custody in Norfolk, Virginia, for violating the trade embargo. On 4 January the RC *Eagle* (Captain Frederick Lee) escorted merchant vessels out of New York Harbor. Two weeks later Captain John Cahoone of the RC *Vigilant* petitioned the U.S. Congress for compensation from his capture of the armed British privateer *Dart*. On 22 January the RC *Active* (Captain Caleb Brewster) stopped a suspicious vessel sailing out of Sandy Hook, New Jersey, headed for England. The cutter boarding party discovered illegal documents, men with no passports in the process of destroying papers, bills and orders indicating that contraband supplies had been provided to British forces, and other evidence of lawbreaking. Nonetheless, a Norfolk, Virginia, newspaper criticized revenue cutter crews and customs officials for alleged overzealousness in embargo enforcement, and for applying customs duties to local merchant vessels. On 22 February the Revenue Cutter *Income* (Master Danielle Elliot), with the assistance of local militia, exchanged gunfire with a British privateer off the Maine coast. The British sailors suffered three casualties, and their cannon fire caused no Revenue Marine casualties. The privateer vessel escaped to Halifax, Nova Scotia.

The selected chronicle of missions and incidents outlined in this chapter convey the extraordinary range of domestic law enforcement and national defense responsibilities the U.S. Revenue Cutter Service and U.S. Customs and U.S. Treasury officials performed against the most powerful naval and military forces in the world. The creativity and courage exhibited by the Revenue Marine, often in coordinated missions with the U.S. Navy, U.S. Marines, U.S. Army, and state militias, established a storied legacy of service that honored the Revenue Marine.

A revenue boat on Lake Champlain in northern New York State was nearly captured during the spring thaw by a British flotilla on 15 March 1814, according to the *Plattsburgh Republican* newspaper. The complexities and politics of policing domestic commerce were illustrated by the instructions from the U.S. Treasury to customs collectors two weeks later. American fishing boats were thereafter exempted from revenue cutter inspections and the provisions of the embargo act of 1813. Meanwhile, the New Orleans customs collector informed the Treasury Department that same month of a violent confrontation that occurred between smugglers and Treasury inspectors and revenue officers.

The politics of embargo enforcement led to an Act of Congress being passed on April 14 that repealed the 1813 embargo so that U.S. cargo vessels could legally leave Atlantic ports. On April 14 Congress recognized the contributions revenue cutter officers and men made to U.S. Navy missions by allowing federal pensions to be awarded to revenue cutter personnel wounded in action. On 11 June 1814 the U.S. Lighthouse Service got into the act when the Scituate, Massachusetts, lighthouse keeper Simeon Bates fired his signal cannon at patrolling Royal Navy boats.

Another successful U.S. Revenue Marine–U.S. Navy mission occurred on July 14 when the RC *New Hampshire* (Captain Nathaniel Kennard) and a USN gunboat captured a Royal Navy supply tender assigned to a British warship and took ten enlisted men and three officers prisoner. On 6 September 1814 Treasury Secretary George W. Campbell received a request from officials in the Delaware Bay area to allow the cutter *General Greene* (Captain Joseph

Sawyer) to be used for intelligence reconnaissance on Royal Navy blockading vessels.

The Revenue cutter *Mercury* (Captain David Wallace) captured a Royal Navy tender that grounded on a North Carolina sandbar on 12 November 1814, taking as prisoners of war a British officer and seven enlisted personnel. On Christmas Eve 1814, Secretary Dallas notified the New Bedford, Massachusetts, collector of customs to interdict domestic commercial vessels that were engaging in "illicit intercourse with the Enemy."[26] That evening, diplomats from the United States and the United Kingdom signed the Treaty of Ghent in Belgium. On 15 January 1815 the 18-gun British warship HMS *Favorite* sailed into New York City Harbor with a copy of the treaty for U.S. officials to present to President James Madison and the United States Congress. On 16 February 1815 President Madison signed the Treaty of Ghent.

On 12 January 1815 Captain Caleb Brewster of the RC *Active* had received orders from American military officials to contact and notify commanders of British Royal Navy squadrons that the war had ended. In the first week of March, Congress repealed the non-importation and embargo acts. Secretary Alexander Dallas sent circular letters with instructions to all U.S. Customs collectors that the war had ended. Cutter officers and crews were to be "recommended for their vigilance, activity, skill and good conduct,"[27] and they were ordered to take "immediate measures ... for restoring the light-houses, piers, buoys, and beacons, within your district and jurisdiction, to the state in which they were before the war."[28]

On 18 May 1815 Captain Caleb Brewster of the RC *Active* sailed from Sandy Hook, New Jersey, to New York City with documents from the USS *Constitution* for presentation to the Department of the U.S. Navy in Washington, D.C. Within months of the termination of wartime hostilities, Secretary Dallas directed customs collectors to initiate the building of cutters to replace the vessels lost, captured, or damaged during the war. The designated cutters and ports were *Search* (Boston, Massachusetts), *Eagle* (New Haven, Connecticut), *Dallas* (Savannah, Georgia), *Detector* (Portland, Maine), *Surprise* (Norfolk, Virginia), *Gallatin* (Charleston, South Carolina), and *Active* (New York).[29] Cutter captains were directed to begin peacetime patrols off American shores.[30]

6

Revenue Marine Missions
in the War of 1812

Dr. William H. Thiesen, the Atlantic Area Historian of the U.S. Coast Guard, has chronicled the missions and significance of the U.S. Revenue Cutter Service in the War of 1812. The role of the Revenue Marine in the war can be better understood by reviewing Thiesen's analysis.[1] The contributions of the Revenue Marine in America's "Second War of Independence," were significant enough for Thiesen to conclude that the "cuttermen rose to the occasion ... and ... forged the identity of the U.S. Revenue Cutter Service."[2]

Dr. Thiesen traced cutter missions against the British Royal Navy on the East Coast, Chesapeake Bay, Gulf of Mexico, and Mississippi River. His geographic survey included smuggler and pirate dens in the Caribbean and maritime activities on the isolated Great Lakes and Lake Champlain. He discussed the cutter officers and crews and Customs and Treasury officials who confronted the global reach of the British Royal Navy and domestic enemies. Thiesen deftly enriched the Revenue Cutter and Coast Guard legacies with stories of courage in combat, while at the same time recounting how the Revenue Marine continued its traditional responsibilities of law enforcement and rescue missions on stormy waters.

Revenue Cutter Service victories over British naval and privateer vessels were significant. Captured enemy vessels had to be escorted into American ports where court adjudication brought prize money to brave revenue cutter crews. Unfortunately, the combat missions included the British capture of some cutters and smaller revenue boats, the loss of cutters at sea, and the casualties and capture of cuttermen, and their confinement in British ports and on prison ships.[3] Revenue Marine enforcement of domestic embargo acts and embargoes led to seizures of American vessels and contraband. Theisen describes cutter successes like the RC *Jefferson* and its capture of three Royal Navy barges (11

April 1813). An American newspaper predicted such seizures would indeed "embarrass the enemy not a little." The June 1812 capture by the *Jefferson* of the British schooner *Patriot*, "aka Prize No. 1," was the first American capture of the War of 1812.[4] "As they would in future conflicts," Thiesen asserted, "revenue cutters served as the U.S. military's 'tip of the spear' with the war's first captures."[5] The RC *James Madison* joined the RC *Jefferson* in successful early action against the Royal Navy by capturing a British brig on 23 July 1812.

The revenue cutters were in the front lines against British barges and warships as they convoyed merchant vessels and patrolled inland and coastal U.S. waters. The cutters gathered maritime intelligence on enemy vessels as they sailed out of American ports, and U.S. Navy ships patrolled further out into the deep blue water. The revenue cutters protected shipping, enforced quarantine and trade regulations, did shallow (brown) water combat and surveillance, and supported lighthouse (aids to navigation) responsibilities.

The sacrifices of revenue cutter crews in time of war are well illustrated in Thiesen's research: 90 revenue crew members were captured as prisoners of war and sent by the British to Halifax, Nova Scotia, in British Canada and to prison-ship hulks in the United Kingdom. One of the prisoners was an African American, Beloner Pault, from the RC *Madison*. A former prisoner of war of the British claimed that up to 250 cuttermen were held in harsh, starvation conditions under English custody. Thiesen's research revealed the overall costs paid by the brave mariners of the Revenue Marine in the War of 1812. In addition to enduring the casualties of war and other seafaring dangers, six out of fourteen saltwater revenue cutters were lost, as well as revenue boats on the Great Lakes.[6] The officers and men of the Revenue Marine also had to be thoroughly familiar with the trade restrictions and fine points of the law they were charged with enforcing on foreign and domestic vessels because American shipping companies and merchant captains would frequently challenge seizures of ships and contraband, crew detentions, and forfeitures they considered illegal.[7]

William Thiesen described the exigencies of naval combat. The maritime historian pointed out the historic significance of the battle Captain John Cahoone and his crew on the RC *Vigilant* waged against the British privateer *Dart* in October 1813. After shooting broadsides at the enemy vessel, cuttermen boarded the privateer that had captured more than 20 American commercial vessels. Thiesen concluded that the incident was "one of the most impressive captures of an enemy ship by a revenue cutter" and "the last known use of an armed boarding party by a revenue cutter during the Age of Sail."[8]

The intelligence and reconnaissance missions performed against British forces were significant and innovative contributions of the Revenue Marine.

The Service tracked enemy movements, located British privateers, provided information to U.S. Navy vessels, transported civilian and military officials, conveyed government and military documents, and kept civilian, customs, military and naval officials updated on events and enemy operations. Revenue cutter crews also gathered information from observant fishing boat crews. The RC *General Greene* monitored the Delaware Bay region and reported the location and numbers of British ships and troop landings and the supplying of British naval and military forces and apprehended American vessels and crews that sold supplies to the enemy.[9] Dr. Thiesen analyzed the significance of the Revenue Marine in the war and the impact of the missions on subsequent maritime history: "During the War of 1812, the revenue cutters adopted new missions. Cutter operations would forever include their previous peacetime responsibilities and their new wartime roles; thereby cementing the core missions the U.S. Coast Guard today."[10] To provide a better understanding of the role and significance of the United States Coast Guard in the War of 1812, Thiesen began his history with an analysis, "Revenue Cutter Operations and Core of Coast Guard Missions." Thiesen titled his 15-page overview "Cementing Coast Guard Core Missions: Revenue Cutter Operations in the War of 1812." This compelling and thorough history is on file with the U.S. Coast Guard Historian's Office at U.S. Coast Guard Headquarters in Washington, D.C.[11]

In contemporary times, the U.S. Coast Guard is responsible for numerous humanitarian, law enforcement, national security, and defense missions. Several of these missions can be traced to the U.S. Revenue Marine and the Revenue Cutter Service at its founding in 1790 under Alexander Hamilton. The Revenue Cutter Service is a "predecessor agency" of today's United States Coast Guard. In the first two decades of the Revenue Cutter Service, the fleet initially numbered ten and then twenty revenue cutters. During the War of 1812 the U.S. Revenue Marine cutter fleet and crews performed the primary functions of guarding American commercial vessels, protecting national maritime commerce, and enforcing trade laws, embargo acts and various commercial regulations, as well as assuming the additional duties of combat and national defense missions with U.S. military forces and the United States Navy. Initially, the revenue cutter service did not have an official name but was referred to by federal officials as "the cutters" or "A System of Cutters." The Continental Congress had disbanded the Revolutionary War navy, so between 1790 and the refinancing and launching of the new federal United States Navy in 1798, the cutters were the only federal navy and can accurately be described as the oldest continuously operating seagoing service of the United States.

The cutters enforced and collected customs duties on commercial vessels because in that historical period before the existence of federal taxes, customs and tariff revenue, and federal land sales were the sources of income for the federal government and therefore the primary sources of federal expenditures and the budget. The U.S. Revenue Cutter Service (USRCS) saved lives and property at sea and conducted surveying and mapping missions. The USRCS boarded commercial vessels that imported and exported goods into and out of U.S. ports. Ship documents and papers had to be checked. Cargoes were sealed, contraband and law breaking vessels were seized, and anti-smuggling laws were enforced. Revenue cutter crews searched for and apprehended smugglers and their cargoes in interior, coastal, and port waters, and out to sea, as well as along isolated shore locations. While busy with those duties, Revenue Marine cutters and boats transported crews and supplies to lighthouses as part of what would later be called aids to navigation missions.[12] American smugglers were an experienced and skilled lot. They had plied their trade since the British Colonial and Revolutionary War periods to counter the trade and tax policies of the United Kingdom and challenged the laws of the federal government of the United States.

The Revenue Marine got its historic start, and fortuitous naval combat training, prior to 1812, in the Quasi-War with France in 1799–1800. The Franco-British wars brought the naval and privateer vessels of the two enemy nations into the Atlantic to prey on American merchant ships. The well-armed revenue cutters operated under the fledgling U.S. Navy and in fact constituted about one-third of the American naval fleet. After 1800 those European naval powers continued to attack American merchant ships and violate the sovereign rights of the United States on the high seas.

Presidents Thomas Jefferson and James Madison and Congress tried to preserve American sovereignty and neutrality with the passage of trade restrictions in a series of laws entitled the (1) Non-Importation Act of 1806, (2) Embargo Acts of 1807–1808, (3) Enforcement Act (1809), (4) Macon's Bill Number 2 (1810), and (5) Non-Intercourse Acts of 1809–1810. As would happen in future U.S. Coast Guard enforcement activities, the Revenue Cutter Service incurred the animosity of the public when it enforced the mandates of Congress because the laws and their enforcement caused the unemployment of American laborers and sailors and financial losses to shipping companies and shipbuilders. The acts also caused British-American conflicts and would eventually be repealed.[13]

On 18 June 1812 Congress passed, and President James Madison signed, a declaration of war against the United Kingdom. Great Britain was the world's strongest naval and military power with more than 600 Royal Navy vessels compared to America's minuscule fleet of 14 revenue cutters with an additional flotilla of small revenue boats and only 16 U.S. navy ships and auxiliary barges.

The U.S. Revenue Marine and its cutter fleet were under the control of the various port customs collectors and the U.S. Treasury Department. Albert Gallatin, the competent Treasury secretary, wrote a circular letter to all of the customs collectors, advising and instructing them as follows: "Sir, I hasten to inform you that War this day has been declared against Great Britain." Then he ordered East Coast revenue cutters to sail out to sea to inform officers and crews of seagoing vessels of the U.S. Navy that a state of war now existed between the United Kingdom and the United States. The available American fleet of revenue and navy warships, barges, and gunboats were now required to prepare to confront British Royal Navy squadrons and patrols in shallow (brown) and deep (blue) water regions and protect America's shores, ports, and commercial vessels.[14]

We have previously covered some of the major mission and combat incidents the U.S. revenue cutters were involved in during the

Treasury Secretary Albert Gallatin dispatched a circular to customs collectors in Atlantic Coast ports advising them that war had been declared against Great Britain. Gallatin ordered cutter commanders to inform U.S. Navy vessels in those ports that war had commenced.

War of 1812 and will not repeat all of those details in this chapter. However, we will be referring to several significant incidents from the perspective of William H. Thiesen's analysis of key maritime events in the history of the U.S. Revenue Cutter Service.

The vulnerable United States border between the New England states and British Canada was patrolled by several cutters, including the venerable RC *Commodore Barry* (Master Daniel Elliott) in the Passamaquoddy District of Maine along the Canadian border. From that maritime region, the *Commodore Barry* seized five domestic smuggling vessels between June 27 and August 2, 1812, and escorted prize ships to Portland, Maine, for local court adjudication, and in response to the charges of carrying illegal cargoes to supply British forces.

Cutter Master George Brooks outfitted the RC *James Madison* and trained his crew to challenge the presence, manpower, and ordnance of formidable American privateer vessels. The *James Madison*, built in 1807, sailed out of Baltimore, Maryland, and then was assigned in 1809 to Savannah, Georgia. The 86-foot cutter was 22 feet wide, and suitable for oceangoing service. After an eight-hour pursuit in July 1812 the *James Madison* captured the 300-ton armed

British merchantman *Shamrock,* a vessel carrying six deck cannon, a crew complement of 16, and a hold filled with arms and ammunition.

Master Daniel McNeill, a former U.S. Navy captain, commanded the RC *Gallatin.* McNeill and his Revenue Marine sailors captured several British merchant ships in 1812, and escorted them to Charleston, South Carolina, for adjudication. During the summer of 1812 the RC *Commodore Barry* crew again distinguished the U.S. Revenue Marine by challenging a squadron of British warships off the Maine-Canada coast that consisted of 4 frigates carrying 38, 36, 18, and 12 guns respectively. The battle was the first confrontation in the war between the Royal Navy and a revenue cutter.

In early August of 1812 Captain Elliott of the RC *Commodore Barry* formulated a defensive plan of action against an approaching patrol of Royal Navy vessels. Elliott anchored next to the well-armed American privateer *Madison,* not to be confused with the RC *James Madison.* The site was in the harbor of Little River, Maine. The American seamen beached the two ships and set up protected shore batteries from the deck guns of the American vessels.

On the 3rd day of August five armed British barges disembarked from His Majesty's squadron, and 250 officers and men attacked the American sailors. The British suffered heavy casualties but emerged victorious. Most of the American sailors escaped into the adjacent forest, but three cuttermen were taken into custody as prisoners of war and transported to a British prison in Halifax, Nova Scotia. The British captured the RC *Commodore Barry,* renamed it the *Brunswicker,* and sailed the vessel out of St. John's in New Brunswick on missions to protect British merchant ships from American privateers and revenue cutters.[15]

Due to the uncertainties and fog of war, the heroic crew on the combat cutter *James Madison* met an ignominious fate. The revenue cutter cruised out of the port of Savannah, Georgia, on 13 August 1812, teamed with the American privateers *Spencer* and *Paul Jones* in pursuit of British merchant ships. On 22 August, in the dark of night, the RC *James Madis*on (Master George Brooks) found and attacked a ship in a British merchant convoy. Unfortunately, the target turned out to be the 32-gun HMS *Barbados* that Brooks mistook for a large enemy merchant vessel. After firing his guns and belatedly terminating his boarding attempt, Brooks sailed from the combat scene in a seven-hour chase that resulted in the capture of the cutter and crew by the *Barbadoes.* The HMS *Polyphemus,* a Royal Navy combat ship-of-the-line, joined the *Barbadoes,* put a Royal Navy crew onboard the cutter, and sailed it to Britain. In 1813, the former RC *James Madison* was sold to a nobleman in Northern Ireland who converted the ex-cutter to an armed yacht. The captured cuttermen were declared prisoners of war, and the paroled cutter officers arrived by ship at New York in

November. The nine enlisted cuttermen were split up into cohorts and differentially dispersed to Halifax in Canada, Boston, Massachusetts, and Chatham, England, for imprisonment.[16]

The new and varied responsibilities of the U.S. Revenue Cutter Service in the War of 1812 led to logistical conflicts and priority disputes. Law enforcement, not naval combat, was still the primary function of the Service, especially given the wartime needs of increased federal revenues and expenditures to address the British blockade and the threats posed by the Royal Navy to American maritime commerce. In the Quasi-War with France (1799–1800), the revenue cutters operated with, and under the jurisdiction of, the U.S. Navy and the War Department. In the War of 1812 the cutter commanders and crews were issued orders from the U.S. Treasury Department and the local port customs collectors. Treasury officials "did not sanction," asserted historian William H. Thiesen, "high seas revenue cutter combat operations."[17]

Cutter crews were still obligated during the war to enforce trade and tariff laws and to guard and escort American merchant vessels. The RC *Thomas Jefferson* protected U.S. commercial routes and vessels, and enforced maritime laws out of the ports of Norfolk, Virginia, and Savannah, Georgia. The *Jefferson*, like other cutters, seized British and American merchant vessels. The RC *Gallatin* performed its law enforcement duties out of Charleston, South Carolina and Norfolk, Virginia. During the course of the war, the cutters were often under the command of more than one master. The skilled and courageous revenue mariners seized American and British merchantmen, and in some instances even captured Royal Navy ships and barges. The *Gallatin* seized the 12-gun HMS *Whiting* on 12 August 1812. Federal officials ordered the release of the *Whiting*, however, because the British warship was performing a diplomatic mission when seized. The RC *Eagle* captured American and British vessels carrying illegal cargoes

Captain Frederick Lee commanded the RC *Eagle* and battled the 18-gun British brig HMS *Dispatch* in the War of 1812 until grounding off Long Island by the port of New York City.

and directed those ships to New London and New Haven, Connecticut, for judicial decisions. In a several-month period in 1812 and 1813, the RC *Eagle* seized and sent into ports for adjudication two British and three American brigs.[18]

Revenue cutters were not placed directly under U.S. Navy and War Department control during the war but did engage in naval combat against Royal Navy vessels and armed British privateers and commercial vessels. The revenue cutters shared intelligence with U.S. Navy ships and authorities, and cooperated with U.S. Navy warships and flotillas. The U.S. Congress, in recognition of the significance of the joint U.S. Revenue Cutter Service and U.S. Navy missions, passed legislation in 1814 that would provided naval pensions to revenue cutter seamen and commissioned officers that were wounded during naval combat operations.[19]

The RC *New Hampshire*, commissioned in 1803, operated out of Portsmouth, New Hampshire. In 1813, the *New Hampshire* (Captain Nathaniel Kennard) patrolled with Commodore Isaac Hull (U.S. Navy), and his flotilla of two gunboats, and the USS *Rattlesnake* and USS *Enterprise*. In November, the flotilla and the RC *New Hampshire* sailed out to sea in search of Royal Navy warships but encountered none. In 1814 Captain Kennard and a U.S. Navy gunboat captured the armed tender boat associated with the HMS *Tenedos*, a formidable 38-gun frigate. The British armed tender had captured an American commercial vessel, and the *New Hampshire* captured 3 British officers and 10 enlisted seamen.

Between 1812 and 1814, the cutters *Active* and *General Greene* patrolled with U.S. Navy vessels, boarded contraband vessels, and broke through a Royal Navy blockading squadron in support of Commodore Stephen Decatur (U.S. Navy) and his Thames River (Connecticut) flotilla. The flotilla included the USS *Hornet*, USS *United States*, and USS *Macedonian*. The RC *Active* provided protection and delivered reports, messages, and intelligence to and from Commodore Decatur and to other military and naval officials.

Master Daniel Elliott personified the skill and courage of cutter commanders in the war. Elliott was now in command of the small revenue boat *Income* out of the port of Machias in Maine. In February 1814, Elliott came in contact with a large, well-armed British privateer and purposely ran the *Income* ashore to allow his crew to take defensive cover in a cove, using defensive small arms that likely included flintlock pistols used by cutter crewmen in the War of 1812. An armed British landing force rowed from the privateer to shore to capture the revenue boat *Income*. Supported by local militia, Elliott and his crew of Revenue Marines inflicted heavy casualties upon the British attackers, forcing them back to their vessel. The following month, Master Elliott captured a British prize ship and brought it into the port of Machias for court adjudication[20] and potential prize money.

Graphic examples of the successful intelligence-gathering missions provided by the U.S. Revenue Cutter Service in the war would include surveillance and reporting on Royal Navy movements, the location and activities of British and American privateers and merchant ships, and the sharing of gathered information with military, naval and civilian officials, customs collectors, and newspapers. Captain Caleb Brewster, the commander of the RC *Active*, was a particularly experienced and successful intelligence operative. Brewster gained his initial experience as a member of a spy ring during the Revolutionary War in the service of General George Washington. Throughout the War of 1812 Brewster continued his courageous missions, conducting surveillance in small boats, often under the shield of inclement weather, against British warships off Long Island and outbound from New York Harbor up to 10 miles out to sea. Captain Brewster forwarded information about British positions, operations, and logistics to Commodore Stephen Decatur and his U.S. Navy squadron, and to civilian, military, naval, and customs officials ashore.[21]

The fast, shallow-draft cutters were called upon to perform their usual law enforcement and rescue missions and carry and deliver military and naval documents, treaties, and other official government papers. The U.S. Army general in command of New York assigned the RC *Active* the responsibility of transporting documents signed by James Monroe, the U.S. secretary of war, to inform the Royal Navy blockade squadron that the 1815 Treaty of Ghent had been signed and ratified and that the war had ended. Revenue cutters had been transporting government documents and officials throughout the war. In August of 1812 the RC *Diligence* regularly carried a U.S. Army general and his staff from their North Carolina command center in the port area of Wilmington to meet with other regional commanders. In that same year, General William Hull, the Michigan territorial governor, utilized a revenue cutter in the Great Lakes watershed region prior to his ignominious military defeat at the hands of the military forces of the United Kingdom. The diversity of cutter assignments is further illustrated by duties assigned to the RC *Mercury*. In July of 1813 the cutter transported U.S. military officers along the strategic and treacherous string of islands of the Outer Banks off the Virginia and North Carolina coasts where military engineers and cartographers surveyed the region to determine the best locations for the construction of forts and landing sites.[22]

With the Treaty of Ghent (1815), the era of wind-powered, wooden-hulled warships that were maneuvered in close-in to deliver and receive broadsides was coming to an end. The necessity of warship commanders to find and hold the "weather gage" to position their vessels upwind of the enemy, who then could not sail into the wind to respond,[23] was nearly over. Yet, during the Civil War, naval vessels would be powered by sail or a combination of sail and coal-fired steam power. But in the War of 1812, the wind-powered sailing ships proved their utility.

The versatility and adaptability of revenue cutters in coastal and high-seas missions were first established during the Quasi-War with France and then in the War of 1812. William Thiesen concluded that the two wars "solidified the cutters' law enforcement and defense readiness roles, and added new missions, including port and coastal security, brown-water combat operations, intelligence gathering, and new naval support missions."[24]

That important mission legacy and adaptability has been continued to the present day, long after the U.S. Revenue Cutter Service and the U.S. Life-Saving Service were merged to form the U.S. Coast Guard in 1915 in the era of World War I.

7

Pirates, Slaves
and Seminole Indians

The domestic and national defense missions and responsibilities the U.S. Revenue Cutter Service faced in the late 18th and early 19th centuries were diverse and challenging. Revenue Marine missions ranged from the enforcement of tariff and trade regulations, ship boarding, protection of maritime commerce, and saving lives and property in the vast American maritime domain to scientific exploration and coastal mapping, supporting the U.S. Navy in national defense and combat missions, and serving American lighthouses and personnel by providing transportation, supply, and inspection functions. Add to those duties the battles the maritime service waged against piracy in the Caribbean, Gulf of Mexico, and off the Florida coast and the Seminole Indian Wars waged between the War of 1812 and the Civil War. The pirates had maritime vessels and plundered American merchant vessels, so Revenue Marine involvement was understandable. But it stretches the mind to understand how the cuttermen were required to confront the land forces and warriors of the proud Seminole Indian nation on the beaches, backwaters, and swamps of the Florida maritime domain.

The U.S. Navy initiated operations against marauding pirates in the Gulf of Mexico and Caribbean Seas, but their ships-of-the-line combat vessels were too large and ship drafts too deep to be effective in the pursuit of small pirate vessels between islands and into shoal waters and humid bayous. The Navy did purchase a few smaller schooners and a side-wheel paddle steamer with which to pursue pirates[1] and achieved some success, but the U.S. Revenue Cutter Service had the vessels, mission experience, seamanship skills, and traditions that made that naval service better suited to the task.

The revenue cutters *Louisiana* and *Alabama* were 50-ton vessels stationed in the Gulf of Mexico. The cutters chased slave ships and pirates, gaining newspaper notoriety from those missions. The 20 October 1819 issue of the *New York Post* placed a reporter onboard the RC *Alabama* who covered the battle between the two cutters and the pirate vessel *Bravo*. The *Bravo* was owned by the notorious pirate Jean Lafitte and, in this instance, commanded by the pirate Jean Lafarge. The confrontation occurred off the island of Hispaniola and the smaller Tortugas. The RC *Louisiana* (Captain Loomis) fired a shot in front of one of the three suspect vessels in the chase, boarded it, and discovered that the passengers on board were victims of robbery at the hands of the pirates. The captives were then transferred to safety aboard the accompanying revenue cutter *Alabama*. The *Louisiana* then pursued the second vessel, which hoisted a flag emblazoned with the name "Patriot." Musket fire from the pirate vessel was answered, and the confrontation ended with a broadside from the guns of the *Louisiana*. The enemy musket fire wounded a revenue officer and three crew members on the *Louisiana*. Sailors from the boats of the two revenue cut-

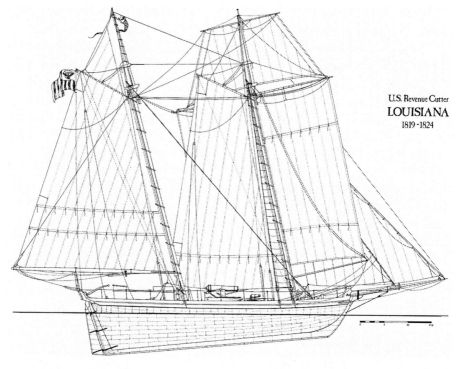

U.S. Revenue Cutter
LOUISIANA
1819-1824

The 52-foot topsail Revenue Marine schooner *Louisiana* carried one deck pivot gun amidships. Constructed in New York City in 1819, the *Louisiana*, while sailing to the homeport of New Orleans, assisted by the RC *Alabama*, engaged in a gun battle and captured the pirate ship *Bravo*, owned by the notorious Jean Lafitte. The cutters released a captured Spanish vessel and took the pirate crew to New Orleans to stand trial.

ters boarded one pirate ship, confronted a crew of about 20, and captured prisoners from other ships. The pirate vessel suffered two crew casualties. The third pirate vessel was boarded and found to be a Spanish ship loaded with male and female passengers and a cargo of baking flour. The cutters took the prize vessels to New Orleans for adjudication. The pirate prisoners were incarcerated and scheduled for court trial.[2]

The naval operations against the pirates, called "buccaneers," were carried out in cooperative missions between the U.S. Revenue Cutter Service, U.S. Navy, and the British Royal Navy.[3] The British-American naval union is interesting and surprising, given the fact that the War of 1812 between the United States and the United Kingdom had concluded within the previous decade. However, common interests in the protection of sovereign rights and maritime commerce influenced the Royal Navy to join forces with the U.S. Navy and the U.S. Revenue Cutter Service.

Breton Island near the mouth of the Mississippi River in Louisiana was the scene of a Revenue Marine raid by the crews of the revenue cutters *Alabama* and *Louisiana*. On 23 May 1820 the *Savannah (GA) Republican* newspaper described how the cutter crewmen landed on the "Pirate Island" and destroyed buildings and supplies with fire. On 29 August the same periodical published a letter written by Captain Jarvis Loomis of the RC *Louisiana* to a New York acquaintance stating how he and his crew captured several pirates, their boats, and contraband goods.

On 25 March 1822 the RC *Alabama* captured three slave-carrying ships that were in violation of the Slave Trade Acts (1794, 1807, 1808) of the United States. Britain had passed similar legislation, and the Royal Navy joined the U.S. Navy and U.S. Revenue Cutter Service in enforcing those laws. On 18 November 1822 the *New York Evening Post* reported that the *Louisiana* had returned to Pensacola, Florida, from patrols along the coasts of Cuba and Florida, during which time five pirate vessels were captured. Two of the vessels burned to the water line, and three boats were delivered as prizes to New Orleans.[4]

The diligence and courage of revenue marine crews and cutters eventually drove pirates and slave ships from the Gulf of Mexico and the Caribbean Sea. The 25 April 1844 *Army and Navy Chronicle* published an article from a Boston-based writer who offered the following description: "The U.S. Revenue Cutter *Hamilton* is undergoing repairs and having new copper put on her bottom." Given the beauty of the cutter and its successful missions, the writer continued, "No wonder she sails swiftly [and] her gallant captain [Josiah Sturgis] is so proud of her."[5] Howard I. Chapelle, the eminent maritime historian and author of the classic *History of American Sailing Ships*, wrote, "Like her sisters, the

Hamilton was a very fine sailer and was used as the supreme test of any schooner claiming a reputation for speed. In races with opium smugglers, fast pilot boats, fishermen, coasters, and yachts, few vessels passed her either on or off the wind."[6]

In the 1830s an emerging and unlikely domestic issue confronted the Revenue Marine: Indian wars. General Andrew Jackson's U.S. Army and militia expeditions against Indians in the southern states and his attempts to move the Native Americans off their hunting lands to locations farther west stimulated violent resistance by the indigenous Seminole Indians and blacks who had fled into the swamps to escape slavery.

Collector of Customs Jesse Willis of St. Marks, a port city located on the west coast of Florida, wrote a letter to the New Orleans customs collector with a request that he send a well-armed revenue cutter to protect the citizenry, and his office facilities, from Seminole warriors. The warriors had allegedly killed white inhabitants and burned buildings and homes. Willis predicted another onslaught by at least "2,000 red men."[7] Customs Collector James Breedlove sent a letter of response from his New Orleans office that arrived in St. Marks two weeks later. Breedlove wrote that he would send the RC *Dallas*, scheduled to arrive in the early months of 1836, to defend St. Marks and Tampa. The *Dallas* would later be recalled to New Orleans and replaced in Florida by the 73-foot RC *Jackson* (Captain Henry D. Hunter), with a complement of 2 officers and 32 enlisted men.

The U.S. Army and U.S. Navy operated with the Revenue Marine in the Seminole Wars in joint operations. The U.S. Revenue Cutter Service contributed ten revenue cutters to those missions. U.S. Navy contractor Samuel Humphreys designed the cutters. The 112-ton Humphrey cutters were sleek, 73-foot vessels with 20-foot beams and 7-foot holds and armed with six guns: 6-pound long guns or 12-pound carronades.[8] The shallow-draft cutters operated in shoal waters larger U.S. Navy frigates could not. The cutters performed diverse and multiple missions, carrying supplies, ordnance, soldiers and marines. Cuttermen went ashore to guard communities and track Indians on sandy shores and into swamps, including the enormous Everglades. It was reported that appreciative settlers and government officials facilitated the awarding of homestead lands to members of the U.S. Revenue Marine who wished to settle in Florida.[9]

Commenting on the contributions the Revenue Cutter Service made to U.S. Army and Navy operations in the Seminole Wars, the commander of the USS *Vandalia* (Captain Thomas T. Webb, USN) wrote the following to Commodore Alexander J. Dallas (USN) in a document dated 12 April 1836: "Their prompt and ready cooperation ... called forth the highest recommendations"[10] from several military and naval commanders involved in that theater of operations. The Revenue Marine contributions to the wars on piracy, slave ships, and the Seminole Wars were significant episodes in American naval history.

The 70-foot sail-powered RC *Louisiana* missions have been discussed, but another cutter named *Louisiana* would be decommissioned shortly after the incident. That RC *Louisiana* engaged in the pirate wars under the command of Captain Angus Frazer of the Revenue Marine. Capt. Frazer confronted two pirate ships off the port of New Orleans. Frazer reported "20 shots were exchanged" and he respectfully concluded, stating "by their superior sailing and night coming on, they made their escape."[11]

The Seminole Wars involved the participation of eight revenue cutters that joined U.S. Navy and U.S. Army operations. The shallow-draft revenue marine vessels attacked Indian warrior concentrations, carried and distributed military and naval ordnance and documents to appropriate authorities, took military personnel to strategic locations, rescued besieged civilians and survivors, harassed enemy warriors at river crossing points, disembarked and recovered military and naval landing forces, and provided mobile artillery support. These missions were carried out along the Florida coastal regions and interior waters.[12]

The suppression of piracy on the high seas and in coastal waters was an essential mission for the Revenue Cutter Service, given its mandate to protect maritime commerce. The work of the revenue cutters *Louisiana* and *Alabama* have been previously chronicled. The RC *Louisiana* took five pirate ships and their crews into custody in 1822 and conducted joint missions with the U.S. Navy and Britain's Royal Navy to diminish and eventually end pirate presence in the Caribbean.[13] The British conducted joint operations with the U.S. Navy and U.S. Revenue Cutter Service because of pragmatic self-interest, given the regional location of several colonial possessions, and the lucrative maritime commerce facilitated by the merchant ships of the United Kingdom.

The commercial interests of the United States required diligent United States Revenue Cutter Service presence in U.S. waters as well. The threats to the collection of essential revenue in compliance with federal tariff laws did not come from just pirate vessels and crews. As president of the United States, former general Andrew Jackson was not inclined to submit to internal threats to public commerce and law and order.

In the antebellum period, from about 1780 to 1860, leading up to the Civil War southern political leaders, writers, and scholars developed "states rights" theories that would lead to doctrines that sought to justify the right of states to nullify federal laws their legislatures deemed discriminatory and economically deleterious. An incident that would prove to be a precursor to the

Civil War occurred in South Carolina. Ironically, the Civil War would begin there years later in Charleston Harbor when Confederate troops fired on Union troops at Fort Sumter.

In the 1830s, President Andrew Jackson responded to South Carolina's declaration of the Doctrine of Nullification and refusal to pay tariffs to the federal government. Jackson directed the U.S. Revenue Cutter Service to send five cutters to Charleston Harbor with orders "to take possession of any vessels arriving from a foreign port, and defend against any attempts to dispossess the Customs Officers of custody." The resolute former general added this warning to his presidential order: "If a single drop of blood shall be shed there in opposition to the laws of the United States, I will hang the first man I can lay my hands on from the first tree I can reach."[14]

Among the Revenue Marine vessels that enforced the law in Charleston Harbor in the Nullification Crisis of 1832–1833, was the 73-foot RC *McLane*,[15] which carried a crew complement of 24, six 9-pound guns on the deck, and various side arms issued to the ship's crew. The *McLane* was named after President Jackson's Treasury secretary, Louis McLane. The Revenue Marine and its successor agency, the United States Coast Guard, have carried out contraband interdiction from 1790 to the present. Since 1794, the USRCS had been tasked with intercepting slave ships on the high seas because slaves were considered illegal contraband. The Revenue Cutter Service was tasked with preventing the importation of new slaves from Africa. From 1794 to the start of the American Civil War in 1861, the cutters captured numerous slave ships and gave hundreds of slaves from those ships their freedom.[16]

Before 1817, there had been clashes between the Seminole Indians and white settlers in the settlements, farms, and plantations of southern Georgia and Spanish Florida. Most of the battles between U.S. military and naval forces and the Seminoles would take place in coastal and central Florida and Lake Okeechobee. Escaped African-American slaves intermarried with the Seminoles. Southern slave owners wanted the federal government to launch military raids to recapture their property, which slaves were considered to be until the Civil War period when their status as human beings and the legality of their freedom as citizens were established by amendments to the U.S. Constitution and supportive judicial decisions.

Many of the Seminoles were Creek Indians in heritage. The former slaves and Indians were well adapted to the humid swamps and semitropical forests. United States naval and military regulars had difficulty coping with that environment and often relied on the skills of friendly Indian scouts to assist them in navigating the region. In clashes between U.S. sailors, marines and soldiers and the Seminole warriors, the often outnumbered Seminoles and blacks inflicted greater casualties on the whites than they suffered. In one instance, the Seminoles killed a boatload of soldiers and their families on the Apalachicola River.

Three years after his 1815 victory over British naval and military forces at New Orleans, General Andrew Jackson invaded Spanish Florida with Creek Indian fighters and 4,000 soldiers. In 1818 the federal government purchased Florida from Spain for $5 million, and Jackson was appointed the territorial governor. In 1828 he campaigned successfully for the presidency of the United States. President Jackson would force Native Americans to trek from their lands into the Trans-Mississippi West with tragic results.

Seminole leaders began to ignore the treaty agreements they believed were signed under coercion and deception. The Seminoles commenced their war against the United States in 1835 under the skilled leadership of Chief Osceola. In 1835 the Seminoles killed 108 soldiers under the command of Major Francis Dade (U.S. Army). One thousand soldiers, led by General Edmund P. Gaines, were isolated in fierce fighting until saved by other U.S. troops. Chief Osceola, who claimed Muskogee Indian heritage, came to speak with General Thomas Jesup (U.S. Army) under a flag of truce in 1837. Osceola was deceived and arrested and died in captivity within a year at Fort Moultrie in Charleston, South Carolina. After several more high-casualty battles against General Jesup and Colonel Zachary Taylor, the Seminoles accepted negotiations in 1842 that allowed those Indians still in Florida to remain. The third and final Seminole war commenced in 1855. After several bloody battles the surviving Seminoles were forced to march across the Mississippi River into unfamiliar western lands.[17]

The Seminole Wars were bitterly fought over an extensive period of time. Historians differ about the specific dates of the chronological periods. One historian starts the wars in 1817. Another historian dates them in the following periods: 1818, 1835–1842, and 1855–1858. Historians contend that the besieged Indians and blacks attacked villages, farms, and plantations. The Seminoles murdered at least one white Indian agent. Approximately 400 blacks fought on the Seminole side until 1838, the year General Jesup decreed freedom from slavery and prosecution for those swamp warriors who joined American forces to fight against the belligerent Seminoles. Among the many American soldiers who fought in the Seminole Wars was Major General Winfield Scott.[18] General Scott had fought in the War of 1812 and the U.S.–Mexican War (1846), and then was appointed the chief of staff of the United States Army. General Scott terminated his career as the commander of the U.S. Army early in the Civil War when age and health issues forced his retirement. But before he retired, General Scott contributed the strategic idea of what came to be called the "Anaconda

Plan" to President Lincoln: the idea of blockading the coasts and rivers of the South using the troop transports and warships of the U.S. Navy and U.S. Revenue Cutter Service to split the Confederacy into regions and gradually deprive the South of the essential imports, exports, and infrastructure needed to sustain the war against the Union.

On 8 May 1858 the United States government officially declared the end of the Seminole Wars.[19] In the first half of the 19th century, the Revenue Cutter Service enforced the Non-Importation of Slaves Acts and freed slaves from bondage on intercepted slave ships. In the piracy conflicts, the Revenue Marine joined forces with the British Royal Navy and U.S. Navy. In the Seminole Wars the U.S. Revenue Marine cutters and crews rescued and protected settlers and landowners and assisted the U.S. Navy and U.S Army in tactics and logistics. The Revenue Marine supported U.S. Army regulars and volunteers with troop deployments and landings and transportation and supplies. The USRCS protected settlements and landowners and strategic river and port locations. Swift and agile revenue cutters provided combat support with small arms and deck guns.

In the Quasi-War with France, the War of 1812, and the Seminole Wars the U.S. Revenue Cutter Service carried out domestic law enforcement, marine safety, surveillance and survey missions, and facilitated the transmission of government dispatches, supplies, and combat support with the U.S. Army, the U.S. Navy, and the land soldiers of the Navy, the courageous and well-trained members of the U.S. Marine Corps. The combat record of experience and achievement would prove to be a significant factor for the U.S. Revenue Marine because they would later serve ably in the Mexican-American War (1846–1848), the Civil War (1861–1865), the policing of Alaskan waters after 1867, and the Spanish-American War (1898).

8

War with Mexico (1846–1848)

The U.S. Revenue Marine had experienced several wars since its creation in 1790: the Quasi-War with France (1798–1800); the War of 1812–1815; the wars against pirates and slavers; and the three Seminole Indian wars (1817–1858). The naval battles were fought in coordination with the United States Navy, and the next war in this chronology would be the War with Mexico (1846–1848). The swift, agile, and relatively shallow-draft revenue cutters would supplement U.S. Navy missions in blockading the coasts of Spanish Cuba and conducting amphibious landings with the U.S. Navy, Marines, and Army.[1]

Shortly after the United States-Mexico War, 2nd Lt. James E. Harrison, in command of the RC *Jefferson Davis*, conducted a joint military operation with the U.S. Army (Company C, Fourth U.S. Infantry) in a mission against belligerent Native Americans in Washington Territory in the Pacific Northwest. In December of 1855 Native American warriors attacked U.S. forces in their camp and killed the commanding officer. Lt. Harrison (USRM) then assumed command and successfully routed the invaders.

In 1858, a decade after the Mexican War, a U.S. Navy force accompanied by the U.S. Revenue Marine sailed to the Latin American nation of Paraguay in response to a quarrel between the United States and that nation over the firing upon of a U.S. Navy vessel by the Paraguay Navy in 1855. One American sailor was killed. The Paraguay government agreed to apologize and pay an indemnity to the family of the seaman. The auxiliary (sail and steam powered) Revenue Cutter *Harriet Lane* (Captain John Faunce, USRM) was ordered to join the 18-ship U.S. naval squadron. Commodore (later Rear Admiral) William Shubrick (USN) would later submit a report to Navy secretary James C. Dobbin that described the contributions made to the mission by the crew of the *Harriet Lane* and the nautical skills of Captain Faunce.[2]

Background issues, politics and diplomatic history must be considered to more fully understand the U.S. military and naval missions in the War with

Mexico. By the 1840s American citizens, politicians and the print media generally endorsed the concept of "Manifest Destiny." That geopolitical concept claimed the United States had an inherent right to expand across North America to Pacific shores. Mexico, newly independent from Spanish colonial status, stood in the path of that destiny. Negotiations, diplomacy, and monetary compensation for Mexico would allow the United States to take control of what is now the southwest part of the United States. If peaceful negotiations and monetary compensation proved unsuccessful, then the United States used military force to achieve its objectives.

President James K. Polk of Tennessee had gained the U.S. presidency during a political campaign that endorsed American expansionism and a pledge to purchase California, its ocean ports, and other Mexican territory. Manifest Destiny premises included the acquisition from Britain of the Oregon Territory in the Pacific Northwest. The United Kingdom agreed to concede part of Oregon pending a monetary settlement with the United States. The Mexican government refused to sell any of its lands to America. The Mexican province of Texas rebelled against Mexico in 1836 and declared its sovereign status as a nation. The U.S. government recognized the secessionist objectives of the Americans in Texas, but the Mexican government did not. President John Tyler had held office immediately prior to the Polk administration. President Tyler and the government of the Texas Republic had agreed to the U.S. annexation of Texas by a joint resolution of Congress in 1845. The Mexican government did not recognize Texas independence or the annexation of Texas by the United States and refused to accept President Polk's offer of $35 million for California and adjacent territories.

President Mariano Paredes deployed Mexican troops across the Rio Grande River, rejecting U.S. claims to the region between the Nueces River in Texas and the Rio Grande River between Texas and Mexico. On 23 April 1846 Paredes declared that a state of defensive war existed between the United States and Mexico. On 25 April U.S. military forces under General Zachery Taylor (U.S. Army) clashed with Mexican troops in the disputed territory between the Rio Grande and Nueces rivers. U.S. forces suffered 16 casualties (wounded and killed), with 47 troops captured by the Mexican army. Polk received the battle news several days after the event. On 11 May 1846 the president, who had already determined to wage war, asked Congress for a declaration of war against Mexico because, as he dubiously claimed, "American blood had been shed on American soil."[3] Congress granted President Polk's request for a declaration of war on 13 May 1846, by a 40–2 Senate vote and a 174–14 vote in the House of Representatives. Opponents of the war declaration insisted that Polk wanted to extend slave territory in the interests of the southern states and that the military clash between troops of the United States and Mexico had occurred on Mexican, not U.S., territory.[4]

President Polk's military strategy was to have America's strong U.S. Navy and U.S. Revenue Marine blockade Mexican ports along the Gulf of Mexico and the California coast, while U.S. Army units were to push the Mexican army south of the declared international boundary of the Rio Grande River and advance upon strategic Mexican towns and cities. Some Mexican military units were well led by professional officers, but overall the military leaders of that nation were not as well trained as were the American officers trained at the U.S. Military Academy at West Point, New York. Mexico had a small coast guard that was no match for U.S. naval ships, crews, and officers. The U.S. naval and military units had the advantage over Mexico of ample general supplies and ordnance and a sophisticated logistical system. President Polk's request for troops from volunteer state militias added several thousand trained soldiers to support the more than 3,00 federal troops on the mission. High numbers of U.S. troops were necessary, given the vast geographic realm that constituted Mexican territory.[5]

The U.S. Army, Navy, Marines and Revenue Marine performed well in the war. Congress supported the services more generously given the fact that the lands acquired from the war made America a continental power. The Treaty of Guadalupe-Hidalgo (1848) gave the United States the territories (later states) of New Mexico, Arizona, Nevada, Colorado, Utah and California. The Rio Grande River was recognized as the international boundary between the United States and Mexico. The United States government agreed to pay Mexico $15 million dollars for Mexican lands and assume an additional $3 million in claims by U.S. citizens against Mexico. President Polk enhanced the image and significance of the president as commander-in-chief in time of war. The expansion of U.S. territory added economic and geopolitical benefits to America but exacerbated the sectional tensions between the slave and free states that would contribute to the Civil War.[6]

Several U.S. military officers distinguished themselves during the war and in military and political life after 1848. Among them were Winfield Scott, Zachary Taylor, Franklin Pierce, George McClellan, and the future Civil War Confederate commanders Robert E. Lee and P.G.T. Beauregard. General-in-chief of the U.S. Army, Winfield Scott, a War of 1812 hero, distinguished himself further in the War with Mexico with his tactical successes, occupation of Mexico City, and use of wooden boats (which would be called landing craft in World War II) to launch ship-to-shore amphibious assaults onto strategic Mexican beaches.

The military historian Stuart Murray claimed General Zachery Taylor

had 4,300 troops, a greater number than the 3,000 troops other historians calculated. But the higher number could include the belated arrival of state militia
volunteers. In early May of the first year of the war General Mariano Arista and
his larger Mexican force were met and defeated by General Taylor because of
Taylor's rapid deployment of mobile artillery and the skilled coordination of
infantry and militia forces. The number of wartime U.S. Army regulars would
eventually total more than 30,00 men, plus 72,000 militia and volunteers,
roughly matching the number of beleaguered Mexican troops.

Among the notable military leaders on both sides in the war with Mexico,
in addition to the military figures mentioned above, was General Antonio
Lopez de Santa Anna on the Mexican side; on the American side were General
Stephen Kearny, who led U.S. troops into Mexico after a long journey from
Fort Leavenworth, Kansas; John C. Fremont; Braxton Bragg; and Ulysses S.
Grant, who would later criticize the war as having been an unjust attack upon
a weaker nation. General Winfield Scott's victorious two-week amphibious
attacks and bombardment from the Gulf of Mexico into several coastal port
cities using naval guns and naval artillery brought ashore paved the way to the
capital of Mexico City and eventually ended the war.

The Mexican-American War cost 17,435 U.S. casualties (dead and wounded,
and from illness and other causes), with 1,733 combat deaths.[7] As mentioned
earlier, United States Naval Academy history professor Craig L. Symonds has
crafted an exemplary narrative of the War with Mexico in his *Historical Atlas
of the U.S. Navy*. The narrative is supplemented with the magnificent maps created by cartographer William J. Clipson. The Gulf Coast maritime theater of
the Mexican War was extensive and complex. The U.S. Navy warships were
often too deep in draft to pass the sandbars of Mexico's Gulf ports. Several warships and transports filled with sailors, soldiers, and marines were able to pass
the sandbars and ascend the rivers and urban areas. The Mexican navy was
small in ship numbers and size and offered little resistance, but strong Mexican
naval shore batteries did significant damage to U.S. Navy vessels. Mexican ports
and cities that were invaded include Tampico, Vera Cruz, Alvarado, and Mexico
City. Preeminent American naval commanders included Commodore David
Connor (USN) and Commodore Matthew C. Perry (USN). The U.S. Marine
occupation of the National Palace in Mexico City provided a theme to the U.S.
Marine Corps service hymn with reference to "the Halls of Montezuma."[8]

Symonds chronicled the military theaters of the Mexican-American War
on the Pacific shores between San Francisco, Los Angeles, and San Diego, and
along the Baja California peninsula to the Mexican port of Mazatlan. The major
American naval and military commanders in those theaters of operations were
American Pacific Squadron commander Commodore John Sloat (USN); Lt.

Col. John C. Fremont (U.S. Army); Brig. Gen. Stephen F. Kearny (U.S. Army); Commodore Robert J. Stockton (USN); and Commodore William B. Shubrick (USN).[9] With the Treaty of Guadalupe-Hidalgo in 1848, the United States acquired California. In 1848 the national Gold Rush occurred after that valuable mineral was discovered near Sutter's Mill, California. From the Mexican War the United States gained the ports and resource realm of the California mainland and the bountiful fisheries of the eastern Pacific Ocean.[10]

The Cutter, a newsletter published by the Foundation for Coast Guard History, featured a synthesis of the missions the U.S. Revenue Marine performed in the Mexican War. In the Summer-Fall 2014 issue of the periodical, the background causes and historical geography of the war between the United States and Mexico were concisely described.

Mexico declared its independence from Spain in 1821, claimed its northern boundary to be the Sabine River in Texas, and granted U.S. citizens the right to settle in Texas and acquire land. The influx of U.S. citizens led to a Texas declaration of independence in 1835. In response, General Santa Anna attacked Americans at the fortress of the Alamo in San Antonio, killing nearly 200 Americans after they mounted a courageous defense. One month later Texas troops led by General Sam Houston defeated Santa Anna's army, took the Mexican general captive, and coerced him into temporarily acknowledging the independence of the new nation of Texas. However, the Mexican government refused to accept the legitimacy of that independence. When the U.S. Congress declared that Texas was thereafter annexed to, and thus a part of, the United States in 1845 the government of Mexico insisted the Nueces River north of the Rio Grande was the boundary between the two nations. General Zachery Taylor (U.S. Army) moved troops across the Nueces in March of 1846, precipitating armed conflict between U.S. and Mexican troops in the disputed territory.[11] In response to those events, the collector of customs at New Orleans, Louisiana, ordered the Revenue Cutter *Woodbury* to sail with a cargo of military supplies for General Zachery Taylor (U.S. Army) and unload the supplies on the beach at Point Isabel. The *Woodbury* then sailed to the Rio Grande River to blockade the port of Matamoros, Mexico.

When the U.S. Congress ceded war powers to President James K. Polk, the Revenue Marine was immediately injected into the Mexican War. Treasury Secretary Robert J. Walker, who would subsequently administer the financing of the war, declared the following: "The Revenue Laws of the U.S. having been extended over the State of Texas, and war with the Republic of Mexico existing, it is deemed advisable to concentrate a number of Revenue vessels between the Rio Grande, Rio del Norte, and the Mississippi."[12] Secretary Walker further ordered the steam cutters *Legare, McLane,* and *Spencer* to duty for combat and to enforce federal revenue regulations. The sailing (wind-powered) revenue cutters *Van Buren, Ewing, Forward,* and *Woodbury* were ordered into the combat theater of operations, and all of the cutters, as Walker decreed, were put "under the direction of the commanding general of the Army of Occupation for the purpose of conveying men, supplies, or intelligence to and from such points as he may direct, and should necessity require, of aiding the forces on board in prosecuting the war."[13] Captain John A. Webster (USRM) was the commander of the RC *Jackson* out of Newport, Rhode Island. Secretary Walker ordered Webster to transfer his command of the *Jackson* and its mission responsibilities to the first lieutenant on the cutter. Captain Webster was ordered to New Orleans to take command of the fleet of cutters stationed in that maritime domain and was then ordered to carry out the instructions of the collector of customs at the port of New Orleans.[14]

The Mexican War period was a significant time in the evolution of U.S. Navy and U.S. Revenue Marine ships. Naval craft changed from wind-powered to coal-fired steam vessels. The RC *McLane* exemplified the transition in her capacity as an auxiliary ship powered by both sail and steam. Sail power was used for long voyages to conserve the coal supply that fired the boilers to create the steam power. In deep (blue) and shallow (brown) water and river combat, steam power allowed greater and more instantaneous maneuverability than was possible under just sail power. The transition from vulnerable and exposed side-wheel steam power to underwater (screw) propeller propulsion brought more complex machinery technology and required complex machine design and maintenance and skilled, knowledgeable officers and enlisted personnel. With that increased complexity, engine breakdowns and boiler explosions were not infrequent dangers. The *McLane* was forced to tie up to the docks during the blockade of the Mexican port of Tabasco because of mechanical breakdowns that would have made the cutter a helpless target had the Mexican Navy been able to launch an attack.[15]

Ulysses S. Grant deemed the war between the United States and Mexico "unjust" in the years after the conflict. At the commencement of the war Congressman Abraham Lincoln stated that the war was "unnecessarily and unconstitutionally commenced by the President."[16] Captain Winslow Foster (USRM), the tough commander of the RC *Woodbury* who served gallantly in the war, later questioned the stated causes and purpose of the war, which he concluded was caused by the "action of Congress in receiving the people of Texas into the Confederacy."[17]

The operational combat theater for the U.S. Navy and U.S. Revenue Marine extended from New Orleans, Louisiana, to the Gulf of Campeche, a maritime region known to U.S. naval forces that patrolled it in the 1830s in order to protect American merchant vessels from pirates and Mexican privateers. In 1838 Captain Farnifold Green (USRM) commanded the RC *Woodbury* in the region to protect merchant ships from belligerent forces. L.H. Duncan of Baltimore, Maryland, constructed the RC *Woodbury* in 1837. The 120-ton wooden-hulled topsail schooner had a crew complement of around 20, and carried four 12-pound guns and one 6-pound gun. The sailing cutter was named for associate Supreme Court justice Levi Woodbury. Captain Winslow Foster (USRM) commanded the *Woodbury* from 1845 and into the Mexican War on patrols from the Mississippi mouth to Corpus Christi, Texas. He carried out Revenue Marine missions of reconnaissance, carrying government dispatches and convoying transport ships for General Zachery Taylor, the commander of the Army of Occupation. Of Captain Winslow, Taylor stated how he "fully relied upon [Foster's] long experience in nautical matters and knowledge of the coast."[18]

Another military commendation honored Captain Foster, this one from Asst. Adjutant General W.W. Bliss from Army of Occupation headquarters and dated 26 March 1846: "Sir: I am directed by the Commanding General [Zachery Taylor, U.S. Army and Army of Occupation] to say, that having executed the service which he required of you, he desires that you will proceed to your proper station…. He takes this occasion to express his thanks for the handsome manner in which you have extended your assistance and that of your vessel to the operations of the Army, and to offer you his best wishes for your health and happiness.[19]" The "proper station" referred to in this letter to Captain Foster was the contested territory between the Rio Grande and Nueces rivers, where the initiation of hostilities between Mexico and the United States commenced.

On 19 May 1846, a week after the U.S. Congress and President Polk declared war on Mexico, Secretary Walker ordered Captain John A. Webster (USRM) to form and lead a squadron of cutters to protect and collect revenue along the U.S. Gulf Coast and assist the U.S. Navy and U.S. Army in the war effort. Webster served in the U.S. Navy before joining the Revenue Marine in a lateral transfer that earned him the rank of captain in 1819. Captain Webster's knowledge of the U.S. Navy made him well suited for his assignment, as had his courageous and competent service in the War of 1812. As expected, he commanded the Gulf Squadron in superior fashion. In 1846, Webster took a leave because of illness, a fever, and turned his command over to Captain Foster. Upon recovering his health, Capt. Webster returned to duty in the USRM, and served until the end of the Civil War in 1865. The Department of the Treasury retired Webster at his highest pay grade in appreciation of his exemplary service[20] to the U.S. Navy, U.S. Revenue Marine, and the United States of America.

In early June of 1846 eleven revenue cutters sailed from their respective port stations with sealed orders that were opened at a specified latitude location and read and discussed by the cutter officers. They were instructed to sail to the port of New Orleans. Two mechanized steam cutters had engine and machinery problems and had to leave the squadron, and the three steamers that completed the journey continued to have engine problems. The sail-powered cutters were more lightly armed than the steam cutters but performed their missions well because of the shallow drafts that got them over sandbars and closer to shore to perform their varied missions.

The steam-powered cutters *Bibb* and *McLane* each had one long gun, an 18-pound deck gun, and one or two pivot guns that shot 18-pound ordnance. The RC *Legare* had lighter guns and one long 18-pound deck gun. The sailing schooners *Wolcott*, *Woodbury*, *Ewing*, and *Van Buren* carried between four and six 12-pound guns. The RC *Forward*, a sail (wind)-powered schooner, carried four 9-pounders. The RC *Morris* had smaller ordnance in number and gun size with 6 six-pounders on deck.[21]

Captain Webster arrived in early August to take command of what would be his squadron flagship, the RC *Ewing*. Stephen H. Evans, the eminent historian of the Revenue Marine and Revenue Cutter Service, referred to Captain Webster as the higher ranking Commodore Webster because he led a squadron from a flag ship and therefore was "the first squadron commodore recorded in the annals of the Revenue-Marine."[22] In addition to carrying out the traditional duty of enforcing revenue laws, Commodore Webster assisted the U.S. Army with his Revenue Marine squadron by doing scouting and reconnaissance, towing boats and disabled vessels, performing blockade functions, and carrying mail, government dispatches, supplies and troops; he was also alleged to have assisted in the suppression of a mutiny onboard a naval vessel. The Revenue Cutters *Legare* and *Ewing* were reported to have unloaded one thousand rifles at Point Isabel that General Zachery Taylor used against Mexican troops at the battles of Buena Vista and Monterey.

Squadron cutters assisted the U.S. Army and the U.S. Navy in support and combat missions. Commodore Matthew C. Perry (USN) commanded a force of naval steamers and two sailing schooners. Perry directed his flotilla from the deck of the USS *Mississippi,* his ship's complement of several hundred U.S. Marines, seamen, and Navy officers; the 6-gun Revenue Cutter *McLane* (Capt. W.A. Howard, USRM), and the 6-gun RC *Forward* (Capt. H.B. Nones, USRM). In October 1846 the naval flotilla anchored in proximity to the Tabasco River. The USS *Mississippi* had to anchor outside the shallow bar. The RC *McLane* lacked the engine power to cross the bar and remained stranded. The RC *Forward* was able to cross the bar at the river mouth and provided

Commodore Perry with artillery fire. Perry transferred to a Navy steamer and Captain Forest put his troops in barges and crossed the bar. The Mexican vessels were captured in the harbor and one Mexican steamer was commandeered as a naval transport for U.S. forces.

Continuing upriver the American flotilla captured a fort and disabled the guns. Observers credited the presence of the U.S. Navy flotilla, 200 Marines, and the guns of the RC *Forward* for influencing the Mexicans manning Fort Acceahappa to abandon their posts. Farther upriver Captain French Forest (USMC) landed with his U.S. Marine detachment and several sailors from the RC *Forward* under the command of Lt. John M. McGowan (USRM) and Lt. W.F. Rogers (USRM). The Mexican troops ashore responded with musket fire, but return fire from the guns of the USS *Vixen* and the RC *Forward* suppressed the enemy fire and forced the town of Tabasco to surrender. Several U.S. Navy personnel were killed or wounded in the exchange prior to the Mexican capitulation.[23] The log of the RC *Forward* described the combat action: "The enemy opened fire on the *Forward* with musketry. We opened up with grape and round shot from three of our guns [9-pounders and a deck pivot gun] with terrible effect for 20 minutes ... (as) a heavy fire of musketry was pouring upon us.... At 12 [noon] we ceased firing by order of Commodore Perry."[24]

The naval flotilla then returned downstream. Commodore Perry ordered the RC *McLane* (which had finally crossed the downstream bar by lightening its load) and RC *Forward* to blockade the port and river. Perry then sailed with his naval vessels and the captured prizes of Mexican boats to the Gulf port of Vera Cruz, near the Yucatan Peninsula. From there he reported in writing to Commodore Connor (USN) of the gallantry and professionalism of his U.S. Navy and U.S. Marine flotilla personnel in the Tabasco mission. Of the contributions of the U.S. Revenue Marine, Commodore Perry graciously said, "I am gratified to bear witness also to the valuable services of the Revenue Schooner *Forward*, in command of Captain Nones, and to the skill and gallantry of her officers and men."[25]

On 2 February 1848, the Treaty of Guadalupe-Hidalgo was signed by representatives of the United States and Mexico, officially ending the war between the United States and Mexico.

9

The Antebellum Period

The "Antebellum Period" refers to the immediate decades before the outbreak of hostilities in the Civil War between the Confederate States and the United States of America. The term refers to the prewar, or "pre-belligerent" period. The Civil War began in 1861 and lasted until 1865. The antebellum period refers to the immediate, or causal or background, period of American history prior to 1861.

Southern states that became the Confederate States of America (CSA) seceded from the United States between 1860 and 1861. The Union never officially recognized the legitimacy of the secession nor did the federal government acknowledge the declared sovereignty of the Confederate States. "CSA" is also used in reference to the "Confederate States Army" it most commonly refers to the states. The Confederate States of America declared its independence and sovereignty as a nation with its first capital at Montgomery, Alabama, and then its second capital at Richmond, Virginia, within 100 miles of the Union capital of Washington, D.C. The Union invaded the CSA to maintain the United States of America and subdue the rebellious Southerners. The CSA, having declared its sovereignty, called the Northern military and naval incursions into the South an invasion and thus referred to the conflict as the War Between the States or the War of Northern Aggression. The United States government perceived the conflict as a war within the nation and therefore designated it a civil war.

The president of the United States during the Civil War was Abraham Lincoln, and the president of the Confederate States of America was Jefferson Davis. The naval vessels of the United States Navy would be referred to as United States Ship (USS) vessels. The naval vessels of the United States Revenue Marine and later the Revenue cutter vessels would be referred to as USRM or USRC or RC vessels. The Confederate States Army would be called the CSA. The Confederate States Navy would be called the CSN. The Union army would be called U.S. Army (USA). The American Navy would be called the USN. And the military arm of the USN was the U.S. Marine Corps (USMC). To understand the causes and administration of the war between the North and the South is a complex but necessary task if one is to appreciate its background, causes, tactics, strategies, results and interpretations The conflict has influenced America and its sectional and socioeconomic geography and politics from the middle of 19th century to the present day.

Jefferson Davis graduated from the U.S. Military Academy and served in the Black Hawk Indian War (1832) and the Mexican War (1846–1848). Between 1845 and 1861 he was a member of the U.S. House of Representatives and served as the U.S. secretary of war and a U.S. senator. The southern states were primarily associated with the Democratic political party and began to secede from the union after the 1860 election of President Lincoln, a Republican. Lincoln was perceived to be a threat to southern political and socioeconomic interests, which included the plantation agricultural system governed by the slaveholding aristocracy. Davis was elected president of the Confederate States of America on 18 February 1861. Despite his military education and achievements Davis did not always make good strategic decisions, because of his alleged sensitivity to criticism and his tendency to punish officers and civilians who challenged him and to protect his favorite military leaders and friends.

Major General Mansfield Lovell (CSA), commander of the strategic New Orleans and lower Mississippi River region, requested more military and naval support in his correct anticipation of an 1862 assault by Union military and naval forces. But Davis instead ordered the Confederate troops and naval gunboats of the river defense fleet to sail farther up the Mississippi River to protect other strategic fortifications. Consequently, in 1862 New Orleans fell to Union naval and military forces. President Davis had ordered Confederate forces transferred out of Louisiana and into Virginia. The governors of Louisiana and Arkansas roundly criticized Davis for leaving their states vulnerable to Union aggression. One governor threatened to secede from the Confederacy if his state was not better defended. Governors began to resist sending their state armies to distant regions, thus leaving the Confederate river cities vulnerable to Union warships and military occupation.[1] General Robert E. Lee, a heroic U.S. military figure before the war, turned down the opportunity to command the Union armies and instead led his native state of Virginia against the Union.

Former New York governor and U.S. senator William H. Seward accepted President Lincoln's request that he assume the responsibilities of secretary of state. In that capacity, Seward managed the administration of diplomatic relations overseas, communications with U.S. ambassadors and counsels, and relationships with foreign diplomatic representatives at their embassies in

Washington, D.C. Seward, like Lincoln, had been initially ambivalent about one of the causes of the Civil War: slavery. Seward thought slavery was a moral evil, but he resisted the clamor for immediate emancipation because of his conviction that such a pronouncement would cause the pro–Union border slave states to secede. Seward favored compensated emancipation over time and believed the territorial westward expansion of the United States would confine and eventually eradicate slavery, given the physical and economic geography of those territories and future states. Seward would later support Lincoln's Emancipation Proclamation, as declared and initiated by the presidential executive order of 1 January 1863.

Seward devoted himself to pressuring foreign ambassadors and the nations they represented, especially Britain and France, to cease their trade and support for the Confederacy. Some European leaders favored the breakup of North America into competing nations. The secretary of state would be best remembered for his fortuitous and far-sighted 1867 purchase of Alaska from tsarist Russia for geostrategic and economic reasons and to further the continental expansionist mission of Manifest Destiny. Secretary Seward was a close friend and advisor to President Lincoln. Seward's loyalty and competence facilitated his controversial control not only of diplomatic affairs but also of internal political, military, economic, national security, and law enforcement issues. William Seward inspected naval vessels, advised the president on political issues, and assisted with the phrasing of speeches and documents. Lincoln did not always accept his ideas, because the lawyer-president, like Seward, was in his own right an exemplary writer and speaker.[2]

On Good Friday, 14 April 1865, the actor John Wilkes Booth assassinated Lincoln. On that same evening, Lewis Powell was assigned to assassinate Seward, who was at his home confined to bed recovering from serious injuries caused by a carriage accident. Powell went to the Seward home and attacked a black servant named William Bell, Seward's son Frederick, and U.S. Army nurse Sgt. George Robinson. The men were stabbed and pistol-whipped trying to save Seward, who, despite his extensive and serious injuries, survived. Powell escaped.[3] A U.S. Army soldier killed Booth a few days later at a Virginia farm. Powell and the other conspirators, including Mary Surratt, the proprietor of the Washington, D.C., boardinghouse where the pro–Confederate plotters roomed and allegedly hatched their plots, would later be arrested, tried, and hanged.

Initially President Lincoln had been ambivalent about slavery, before he announced the Emancipation Proclamation in 1863, and had proclaimed that the dominant mission of the United States was to restore the Union. He had stated that if he could save the Union by maintaining slavery he would do so, or if he could save the Union by abolishing slavery he would do so. He favored compensated emancipation for slaveholders who freed their slaves and also advocated the voluntary colonization of freed and escaped slaves back to Africa. President Lincoln issued the Emancipation Proclamation for moral and diplomatic reasons, the latter driven by his objective to keep the antislavery British from continuing to diplomatically and militarily support the Confederacy.

Slavery and the sectional and economic aspects of that oppressive and violent system were causal factors of the Civil War. Other causes included the different regional (sectional) geographies and political and economic interests of the northern and southern states, and the states rights philosophy of decentralized and regional autonomy. The earliest form of government created by the colonies and later the United States in their quest for freedom from British control was a confederacy supported by the states rights-oriented Articles of Confederation, the first American constitution. The Southern states assumed the legitimacy of that historical precedent. Ironically, the Confederate States quarreled over their regional differences and often resisted, in the name of state sovereignty, many of the orders, demands, and requests imposed upon them by President Jefferson Davis from the Confederate capital at Richmond.

The issues of slavery and abolition exacerbated the conflicts that let to southern secession and the Civil War, but slavery was not the only cause. Most southerners did not, nor would they ever, own slaves. Confederate soldiers and sailors, for the most part, were fighting a defensive war to protect their lands and families. Most Northern sailors and soldiers were fighting to preserve and save the Union. The Border slave states stayed with the Union. Demographic and economic changes further separated the interests of the North and the South. The North was more industrial, the South agricultural. The South bought northern manufactured goods. The North bought southern agricultural commodities such as rice, cotton, tobacco, and sugar. The North was the beneficiary of a large European immigrant population that expanded the labor population. The expanding labor force and economy fueled the growth of industry, the production of manufactured goods, and the infrastructure and technology that put the South at a significant competitive disadvantage during the antebellum period and throughout the war.

The prewar decade from 1850 to the commencement of the Civil War in 1861 culminated in a 25 percent growth rate in the northern farm population and a 75 percent increase in the urban population. New York City reached a population of 800,000. Northern population growth and the westward expansion of the population gave the North greater control of Congress and the legislative process.

The Union was able to out-produce the Confederacy in food and the technology of military and naval warfare, a logistical and tactical advantage accrued with the multiplier effect of increased industrial production, supplies, and transportation. With that, for the purposes of waging war, came an increase in the production of combat vessels, weapons of war, and a more extensive railroad system used for the transport of troops, supplies, animal forage, and weapons of war. The Union could manufacture and purchase the necessary naval vessels with which to divide and blockade the Confederacy at its rivers and ports and attack the naval and military infrastructure of the South.

The advanced technology of war resulted in a death toll of more than 620,000 USA and CSA sailors and soldiers, not to mention the huge numbers of civilian casualties. Members of the armed forces on both sides were wounded in action and killed by accidents and disease including hospital and surgical infections. The impact of the war and postwar periods on the physically and emotionally wounded survivors is beyond calculation. The Union had a population of about 30 million at the start of the war, the South about one-third of that. Of the 10 largest cities in the United States in 1860 the North was home to 9 of them. The South had one: New Orleans, Louisiana.

Political leaders from the North and the South who wanted a political and diplomatic compromise on the sectional issues were blocked by extremists on both sides who would not even consider compensated emancipation for slave owners who freed their slaves. The slaves were termed "contraband" in Union parlance. Attempts at territorial compromise (extending the geographic boundary lines, south of which slavery would be allowed, north of which freedom would prevail) failed. Attempts to restrict the geography of slavery and to grant freedom to slaves seemed to many southerners to violate the Tenth Amendment of the U.S. Constitution, which stated the powers not granted to the federal government or prohibited by that government would be reserved for the states. The "states rights," or state sovereignty issue, was important to that section of the nation that was increasingly perceived as being weak. The failure to resolve the sectional strife led to the secession of the southern states, the formation of the Confederate States of America, and the Civil War of 1861–1865.[4]

Although some historians prefer to simplify the causal factors of the Civil War into just the category of slavery, reducing other alleged causes to simplistic and deceptive analysis, the historian Kenneth Stamp among other historians have published more complex and multifaceted analyses. To some observers, the single-issue slavery causation thesis might constitute a psychological and philosophical attempt to grant a superior moral authority, motivation, and mission to the North. Professor Stamp and other historians have offered multiple, sometimes interconnected, causal factors while conceding that slavery was the most significant cause of the sectional conflicts that precipitated the Civil War.

In their taxonomies, Stamp and other contrarian historians list such factors as the following: Radical Republicans and abolitionists whose moral fervor would not accept compromise solutions; the slave power oligarchs who would not allow open discussion of alternatives or accept adjustments or modifications to their "peculiar institution"; states rights advocates; historical determinists who relied on the precedent of the original Articles of Confederation to rationalize the states rights (confederacy) origins of the United States; economic differences and interests between the North and the South; blundering politicians and agitators; northern and southern reactions and rationalizations about the brutality and immorality of slavery; and a conflict of cultures and regional stereotypes, biases and ethnocentric generalizations.

Religious divisions also contributed to the sectional conflicts and debates. The division of Christian denominations over the rights and wrongs of slavery split the Christian churches and denominations along sectional and theological lines, each side using the Bible for support. Other philosophical premises considered minority and majority rights; aristocratic conflicts and class struggles; the socioeconomic gulf between the slave owners and so-called common people; and the crises of fear and racist ideologies about the social conflicts and violence that might erupt after the manumission of slaves. Some northern farmers, manual laborers, and industrial workers worried that the migration of former slaves into "free soil" regions would cause labor and land competition and violent conflicts. Certainly the element of racism intruded into the northern and southern psyches. Even Abraham Lincoln doubted that white and black social amalgamation should, or would, occur, and he suggested colonizing former slaves into Caribbean and African areas of high black populations.[5] As we have seen, slavery was a significant cause of the Civil War but not the only cause. In fact, some Union soldiers testified that they had not entered the war to free the slaves and if that had been the initially declared objective they would have withdrawn from the military struggle, as some in fact did after the emancipation declaration.

The industrial northern states, and their representatives in Congress, imposed high tariffs on European imports. That forced European industries to raise prices on goods to compensate for the tariff costs, which hurt the economy of the South reliant on the purchase of European industrial goods. Northern tariffs were imposed to raise federal revenues and force the South to buy more northern industrial goods like farm machinery and other manufactured items at the higher prices. The southern states assumed and demanded a "Right of Nullification" to disobey federal laws deleterious to their economic and political interests, but the federal government did not recognize that assertion as being legitimate.

A national presidential election was held in 1860. The new Republican Party and its delegates, after competitive debates and several votes, chose Abraham Lincoln as their candidate. Lincoln was perceived as a moderate (potential compromiser) on the issue of slavery, and he initially concluded that the federal government lacked the authority to interfere with or abolish slavery, a position he would abandon as president when he issued the Emancipation Proclamation in 1863. The Republican platform did, however, threaten southern sectional political and economic interests with its program for free eastern farmlands, railroad subsidies across the nation, and tariff increases. The northern Democrats supported Senator Stephen Douglas for president to succeed President James Buchanan. Southerners rejected Douglas because, in his attempt at sectional compromise, he had created the concept of "Popular Sovereignty," which allowed settlers to vote against (as well as for) slavery in their territories. With four candidates on the ballot Lincoln got a minority of the national popular vote (40 percent) but enough electoral college votes (180) to win the election. In March of 1861 Lincoln assumed the presidency of the United States. South Carolina had seen the political writing on the national wall and had seceded from the Union in December of 1860. In February 1861 other southern states seceded, set up a convention in Montgomery, Alabama (on 9 February), formulated and ratified a constitution, and formed the Confederate States of America. Jefferson Davis of Mississippi was elected president of the CSA. Alexander Stephens of Georgia was named vice president. The first CSA capital was in Montgomery, Alabama, and was later moved to Richmond, Virginia.

Both Lincoln and Davis were born in Kentucky. Lincoln moved north, Davis south. Davis had strong Union ties as a former U.S. Army officer, member of the United States Congress (elected first to the House and then to the Senate), and had been a U.S. secretary of war.[6] After the Confederacy was formed, President Lincoln made it clear that he would fight to maintain the Union and recover federal properties and forts that the CSA claimed, held, or threatened.

Major Robert Anderson (U.S. Army) commanded Fort Sumter in Charleston Harbor. President Jefferson Davis gave General P.G.T. Beauregard permission to fire on Union supply ships and bombard the fort to force its surrender. After a 34-hour bombardment that commenced on 12 April 1861, Major Anderson surrendered. There were no battle casualties on either side, but two U.S. soldiers died when explosives accidently detonated during a final salute to the United States flag. Major Anderson and his troops boarded an American steamship for the journey back to the port of New York. The Civil War had commenced. A Confederate flag flew over Fort Sumter on 14 April 1861.[7]

President Lincoln responded to the start of the war with a call to the states for 75,000 volunteers to serve 90-day enlistments. That enlistment period would be extended when the first battles between USA and CSA troops indicated a long war ahead. Lincoln's call to arms occurred on 15 April 1861, with a presidential precedent dating back to President George Washington's administration and a statute granting the president the authority to call state militia forces into the service of the federal government when lawbreaking occurred by "combinations too powerful to be suppressed"[8] by available federal military and naval forces. The Federal military establishment of regular troops totaled about 16,000 men, a number inadequate to cover the Indian frontier and other inland regions and river and coastal fortifications. Volunteers would have to supplement regular enlisted troops and commissioned officers. Popularly elected officers would lead volunteer regiments. State militias would have to be quickly trained and drilled and then absorbed into Federal army ranks.

Southern militias would be trained and inserted into the regular ranks as well. Jefferson Davis knew military training and tactics, having graduated from the U.S. Military Academy at West Point, New York. Approximately one-third of the West Point-trained officers in the regular Army decided to fight for the Confederate cause. The Confederates were fighting for independence and their home territories. The Union soldiers and sailors would be fighting to save the Union. Slavery was not the burning issue or motivation it would become later in the war. The North would have to plan an offensive expanding war. The South would fight a defensive war. The North had a population exceeding 18 million; the South had about 9 million, a third of which were, in the language of the day, Negro slaves.

General Winfield Scott, the top U.S. Army commander, planned for the training and equipment of the army, and strategized for the necessity of using U.S. Army and U.S. Navy and U.S. Revenue Cutter assets and personnel to conquer interior rivers and Confederate fortifications and initiate a blockade to block Southern exports and imports. That logistical and tactical operation was later called the Anaconda Plan.[9] The Union would have the advantage in possessing, building, and buying boats and warships, and in crew numbers. The U.S. naval forces would come from the experienced U.S. Navy and U.S. Revenue Marine, which was, of course, later called the U.S. Revenue Cutter Service. Both sides in the war offered recruitment and reenlistment bonuses. An early Northern recruiting poster promised future seamen cash prizes, chances for advancement, exciting travel and adventures, and the chance to be heroic patriots. Sailors would fight on interior waters, rivers, coasts, Gulf waters, sounds, and ocean maritime domains.

The training, tactics and logistics of naval warfare were in transition in the Civil War era. Warships were evolving from wooden to iron vessels and from wind-powered sailing ships to coal- and steam-powered vessels propelled by exposed side-wheel paddles, and subsurface screw propellers. The ironclad vessels were used in the navies of both the North and the South. The South even developed a submarine. Ordnance (guns and ammunition) was advancing at a fast and powerful rate, making shipboard battles bloody and dangerous. The South innovated in ingenious ways and built, bought, and purchased vessels at home and in Europe. Steam frigates were wooden vessels rigged with sails to preserve coal supplies on long voyages. Steam power provided agility on coastal and river waters and in close combat. The Union navy initiated the blockading of the extensive waterways and ports of the Confederacy, but blockading increased the diplomatic complexities of the war. Blockades were an act of war against another nation, and the Lincoln administration did not want European nations to perceive the CSA as an independent nation and thus be tempted to support the South for economic and geopolitical reasons.

As was the case in the ground war, the war on the waters was essentially a defensive war for the Confederate States of America, and an offensive war for the United States of America.[10]

10

The Confederate and Union Navies (1861–1865)

Confederate navy secretary Stephen R. Mallory had chaired the U.S. Senate Committee on Naval Affairs as a senator from Florida. Mallory knew that in 1861 he had only 10 naval vessels with which to compete with the formidable U.S. Navy and U.S. Revenue Marine. During the Civil War the USRM would be referred to as the U.S. Revenue Cutter Service (USRCS).

The Confederate States Navy (CSN) would need to buy, construct and capture transport vessels, warships, and merchant ships. The CSA commissioned privateers and naval vessels to prey upon Union commercial vessels. The Confederate army and navy needed shore and ship guns to protect Southern coasts, rivers, and forts and needed to strengthen and supply the hundreds of forts and gun emplacements stationed at harbor entrances and along river and ocean shorelines. The Confederate privateers and naval vessels would need to challenge the Union naval blockade of Southern ports and go out to sea to harass, capture, and sink Union merchant ships and whalers in order to destabilize the Northern economy, raise commercial vessel insurance rates, and draw ships from the formidable Union navy out to sea in search of the Confederate commerce raiders.

The Southern Confederacy was fortunate to have the more than 300 experienced naval officers who left the U.S. Navy to join the CSN. Commander Matthew Fontaine Maury had been a commissioned officer and oceanographer in the USN before he transferred to the CSN. Commander Maury pioneered in the study of the physical geography of the oceans. Applying his creative genius and nautical skills to the Confederate navy, Cmdr. Maury invented the naval "torpedoes" that would later be called "mines." The mines were strategically placed by cable, chain and weights in critical waterways, and they claimed several damaged or sunken U.S. naval vessels. John Mercer Brooke designed the Brooke rifle, a powerful naval cannon that was as effective as the Dahlgren deck gun invented by Brooke's former Union commander Admiral John A. Dahlgren (USN). Dahlgren was the U.S. Navy's formidable artillery and ordnance expert.

Commander Catesby ap Roger Jones (CSN), a former USN lieutenant, commanded the Confederate States Ship (CSS) *Virginia* (former USS *Merrimack*) against the USS *Monitor* in the first naval battle of iron-hulled warships and ironclads, off Hampton Roads, Virginia, on 9 March 1862. (The "ap" letters in Cmdr. Jones's name is a Welsh reference meaning "son of"). Representatives from the CSA purchased commercial vessels at Southern ports and used them as transports and merchantmen or converted them into warships. The Confederates contracted for ships and the new submarines, purchased ships from European shipyards, and commandeered vessels in Southern ports,[1] including a few Revenue Marine vessels taken over by commanders who transferred their allegiance to the Confederate States of America.

Confederate blockade-running ships (raiders) carried exports and imports crucial to the civilian and military needs of the CSA and the war effort. The blockade-runners were fast and sleek and more like civilian yachts than warships or traditional commercial vessels. The funnels (smokestacks) could be retracted to make the ships less visible, especially at night against the horizon, and smokeless coal made the ships even more difficult to distinguish. The South lacked the numbers of skilled laborers and factories needed to turn out iron armor, warships and raiders, but it did manage to construct twenty-one ironclad vessels. Confederate ironclads and torpedo mines caused understandable anxiety among Union blockade crews that patrolled the rivers and harbors in the enemy maritime zones. The innovative Rebel naval architects designed submarines. The brave sailors in the 40-foot, hand-cranked, propeller-driven CSN/CSA *H.L. Hunley* managed to sink the blockade ship USS *Housatonic* (17 February 1864) in the outer reaches of Charleston Harbor in South Carolina. The crew used a detonating spar device attached to the submarine's bow before sinking the *Housatonic,* but the explosion unfortunately sank the *Hunley,* taking the eight submariners to the bottom. Nonetheless, the iron coffin, as some observers called the submarine, is etched in history as the first submarine to sink another warship in battle.[2]

To match the overwhelming industrial, military and naval strength of the Union, the Confederacy used cotton bales in place of armored protection on the decks of ships and boats, attacked bulky Union blockade vessels farther off shore at night with agile small craft; attacked, from water and shore sides, Union naval vessels tied up in ports; and captured several Union vessels. Confederate agents operated across the Great Lakes and in New England out of neutral

British Canada to capture supply ships, intercept railroad shipments, and in failed attempts to liberate Confederate prisoners from Great Lakes prison camps.

Southern commercial raiding vessels operated far out to sea in the Atlantic, Pacific, and Indian oceans and in the Bering Sea and destroyed enough commercial vessels to raise maritime insurance rates. The commerce raiders took civilian prisoners who were eventually released to other vessels at sea or onshore. Especially successful were the CSS *Shenandoah* and CSS *Alabama*, the latter under the command of Captain Raphael Semmes (CSN). The CSS *Alabama* was sunk in 1864 off Cherbourg, France, in the English Channel by the USS *Kearsarge* (Capt. John A. Winslow, USN) after completing a global sailing mission that claimed 65 American merchantmen. The CSS *Shenandoah* ravaged North Pacific whaling ships in the attempt to destabilize the New England economy. Southerners rejoiced over that mission because they attributed the abolitionists of New England with starting the War Between the States. The seamanship skills of the Confederate crew members of the *Shenandoah* were lessons in courage in the face of danger, especially when one considers that most of the Southern sailors were more familiar with a semitropical climate than the cold and ice and storms of the High Latitudes. The raider crews in the North Atlantic and North Pacific had to endure and survive the hazards of ice-covered seas in subzero temperatures and fog, dodging icebergs in their fragile, wooden, copper-bottomed vessels. Some neutral ports allowed the sailors to resupply and repair their ships. After the war the United Kingdom would adhere to the decision of an international claims court and pay damages to the United States in 1872 because the *Alabama* and other Southern raiders and warships had been built in the United Kingdom and sold to the Rebel officials and sailors.[3] The British were therefore accused of prolonging the war and violating the obligations of official declarations of diplomatic neutrality. Commander James D. Bulloch (CSN) was the CSA agent in Liverpool who got ships built and purchased and supplies provided for the CSN, although the U.S. ambassador to Britain (Charles Francis Adams) and the U.S. consul in Liverpool England (Thomas H. Dudley) had tried to thwart Britain's assistance to the Confederates. An international tribunal met in Geneva, Switzerland, from December to September in 1871 and 1872 and awarded the United States $15 million in gold from British coffers as compensation to the United States for Britain's failure to enforce its neutrality laws by providing ships to the Confederacy.[4]

The 220-foot auxiliary (sail- and steam-powered) CSS *Shenandoah* was under the command of Lt. Cmdr. James Iredell Waddell, a graduate of the U.S. Navy Academy at Annapolis, Maryland. Commander Matthew Fontaine Murray

(USN/CSN) gave Waddell the oceanic textbooks and charts he had authored and constructed, which assisted the Southern sailors in their global mission. The globe-trotting crew of the CSS *Shenandoah* returned to a British port in November of 1865, after the end of the Civil War, and the ship was auctioned off. The successful voyages included the capture and burning of American whaling ships as far north as the Bering Sea and coastal Siberia and the capture of 38 merchant ships as prizes. Crews feared capture and trial by Union officials for the crime of piracy, even though the ship carried letters of marquee that were official documents from a sovereign nation that authorized the missions and were intended to avoid piracy charges. The letters of marque allowed ships into neutral ports for supplies and repairs, and for remuneration for the value of the captured prize vessels.[5]

The 220-foot commerce raider CSS *Alabama* (Capt. Raphael Semmes) captured 60 merchant ships in its transoceanic (Atlantic and Caribbean) voyages until the warship was sunk in the English Channel off the coast of France on 19 June 1864 by the USS *Kearsarge* (Capt. John A. Winslow). The *Alabama* was in the port of Cherbourg seeking badly needed repairs after a 22-month voyage, during which the crew captured 65 Union merchant ships and sank the USS *Hatteras* off the Texas coast. The size and complement of the *Kearsarge* was similar to the *Alabama*, with seven deck guns on the *Kearsarge* and 8 on the *Alabama*. The two captains had known each other from their time in the U.S. Navy in the Mexican War of 1846–1848, French authorities had negotiated with the officers of both warships so that the confrontation, witnessed by numerous French and British boat and ship passengers, would occur more than six miles off the coast, outside the 3-mile territorial limit, to ensure naval shells did not land on French territory. This engagement would be the last major naval battle between wooden warships.[6]

The *Kearsarge* battery of guns was superior to the *Alabama* ordnance, its powder fresher, and its crew more practiced, especially with the formidable Dahlgren and pivot guns. Captain Winslow had ordered chains slung over the midsection of the wooden hull and covered it with boards to protect the ship's machinery. Captain Semmes of the *Alabama* would later protest that the chain shield was unfair and that the process had converted the *Kearsarge* into an iron-clad vessel. After the battle the *Kearsarge* rescued 63 wounded and waterborne *Alabama* sailors, but Capt. Semmes escaped capture along with 34 officers and men who were rescued by the English yacht *Deerhound*, which took them across the English Channel to the United Kingdom.

Captain Semmes (CSN) and Captain Winslow (USN) personified the well-trained, versatile, courageous and skilled naval seafarers who risked their lives to master the technology and seamanship required to sail their warships into the jaws of storms and death in the service of their respective causes and nations. Captain Semmes, born in Maryland in 1809, was appointed a U.S. Navy midshipman in 1826, lieutenant in 1837, and commander in 1855, six years before the Civil War. When not on cruises, he studied law and was admitted to the bar (1834). During the Mexican War (1846–1848) he survived after his ship, the USS *Somers,* sank in a gale while on blockade duty off the port of Vera Cruz, Mexico. Semmes wrote two well-received naval histories about his adventures in the Mexican War and then the Civil War: *Ashore and Afloat During the Mexican War,* and *Ashore and Afloat During the War Between the States,* the latter published in 1869. As a result of the battle with the *Kearsarge* off the coast of France in 1864 Semmes lost his ship and 19 sailors, who were either drowned or killed in action. He returned to the United States, was arrested, tried on the charge of treason, and acquitted. Semmes had been incarcerated for 3 months and then was freed. He then moved to Alabama, where he practiced law until his death in 1877.[7]

Captain John A. Winslow (USN) was born in the port city of Wilmington, North Carolina, in 1811, moved to Massachusetts (1825), and was appointed a USN midshipman (1827), then lieutenant (1839), and commander (1855). In the Mexican War, Winslow, like Semmes, lost his ship while on blockade duty. The two navy officers became friends when they served aboard the USS *Cumberland.* In 1861 Winslow was a lighthouse inspector in the Boston area. U.S. Navy officers joined with their U.S. Revenue Marine colleagues in U.S. Lighthouse and U.S. Life-Saving Station inspection duties. Winslow then took command of the USS *Benton* on patrol on the Mississippi River. After recuperating from a shipboard injury, he was promoted to the rank of captain in 1862. President Lincoln came belatedly to the cause of slavery emancipation, and Captain Winslow, treading on the line that forbade commissioned officers from overtly engaging in politics, criticized the administration for not quickly and clearly declaring emancipation a wartime objective.

In 1862 Captain Winslow was placed in command of the USS *Kearsarge.* On that warship, Winslow patrolled the Atlantic Ocean and the Mediterranean Sea between North Africa and southern Europe. The *Kearsarge* also patrolled on the English Channel between France and Great Britain. The mission on these waters was to interdict Confederate raiders. It was on the English Channel patrol that Capt. Winslow encountered and sank the CSS *Alabama* in the historic one-hour battle. His victory earned him a promotion to the rank of commodore. Some naval critics, however, cited Winslow's failure to chase down and fire upon the British yacht *Deerhound* as it carried his former U.S. Navy friend Raphael Semmes and his surviving crew to safety in Britain. After the war Commodore Winslow commanded the Gulf of Mexico Squadron. In 1870 now Rear Admiral Winslow was put in command of the United States Navy Pacific Fleet. Admiral Winslow retired in 1872 and died the following year.[8] Several U.S. Navy ships have been named after Admiral John Ancrum Winslow.

After the Union's Fort Sumter in South Carolina fell to the CSA in April of 1861, President Lincoln initiated the naval blockade of Southern ports. However, with the vast distances and numerous ports involved, the U.S. Navy and U.S. Revenue Marine were initially unprepared to adequately assume and complete these missions. The commanding general of the U.S. Army, Winfield Scott, created the Anaconda Plan, which was designed, in the manner of the huge snake of that name, to crush the Confederacy by using U.S. Navy warships, Revenue Marine cutters, and U.S. Army and U.S. Marine troops to do amphibious and ground operations intended to defeat Confederate troops and naval forces on land and along the innumerable Southern rivers and coastal waters.

The U.S. Navy had about 90 vessels, only half of which could be on station and fit for duty in American waters. Many warships were on overseas sailing missions and would take weeks or months to get back to U.S. waters. The Home Squadron consisted of just eight ships, only four of which were screw steamers; the other four were sailing ships, and all the warships had vulnerable wooden hulls. The Confederate maritime domain consisted of approximately 200 ports and 3,500 miles of coastline stretching from Norfolk, Virginia, on the Atlantic Coast to New Orleans, Louisiana, on the Gulf of Mexico. An effective blockade was estimated to require 600 ships and an enlistment of 18,000 sailors, with more than one-quarter of the commissioned U.S. Navy officers resigning to join the Confederate Navy. U.S. Navy secretary Gideon Welles and Assistant Secretary Gustavus V. Fox worked with federal clerks, government bureaus, and the Revenue Marine to prepare for the war ahead and carry out President Lincoln's "Blockade Proclamation." Naval bureaus and boards were established to build and purchase the hundreds of ships and boats and train the crews necessary to complete the mission. Wooden ships, ironclads, steamers, semisubmersible iron monitors, merchant ships, ferries and tugs were authorized for construction and acquisition.

The U.S. Navy chased blockade-runners, examined ships' papers (the U.S. Revenue Marine, with its shallow draft cutters and boarding and law enforcement experience was especially effective in these missions), destroyed enemy fortifications, purchased supplies, transported civilian and U.S. Army personnel, and hired knowledgeable civilians to advise and guide local operations. Competent, courageous and creative U.S. Navy and U.S. Revenue Marine officers rose to the occasion to plan and execute combat strategies. U.S. Revenue Marine (later the U.S. Revenue Cutter Service) personnel and missions will be referred to in this chapter and more extensively covered in Chapter 12.

A partial list of Union and Confederate naval notables includes Gideon Welles, Gustavus V. Fox, David Farragut, David Dixon Porter, John Winslow, Raphael Semmes, James I. Waddell, Matthew Maury, and John Dahlgren.[9] Gideon Welles, a former Democrat, was a vehement abolitionist who joined Lincoln's post–Whig Republican Party and served as U.S. Navy secretary from 1861 to 1869. Lacking naval experience, he nonetheless made fortuitous and timely wartime decisions about the use of ironclads, steam machinery, heavy ordnance and armed cruisers, as well as accepting the advice and knowledge of the assistant navy secretary and U.S. Navy veteran Gustavus Fox.[10] The Massachusetts-born Gustavus Vasa Fox graduated from the U.S. Naval Academy at Annapolis, Maryland, in 1838, served in the Mexican War, and earned the rank of lieutenant. He resigned his commission in 1856 to go into business. He later would reject Lincoln's wartime offer to command a ship but accepted the specially created position of assistant secretary of the Navy under Secretary Gideon Welles. Fox's leadership in personnel management, engineering decisions, ship contracting and construction, and eventually curriculum modernization at the U.S. Naval Academy officer's school proved invaluable.[11]

David Farragut (USN) was commissioned a midshipman in 1810, served under Commodore David Porter in the War of 1812, and by 1855 had earned his captain stripes. The Virginia native moved north in 1861 when Virginia seceded from the Union. He accepted command of the West Gulf Blockading Squadron from Secretary Welles, with orders to form and train a navy squadron to ascend the Mississippi River, destroy and sail past Fort Jackson and Fort St. Philip, and capture New Orleans with troop transports and a force of 19 mortar boats (under Commander David D. Porter) that fired gigantic shells out of 13-inch–bore cannons in arching patterns over fort walls from distances of up to two miles or more. Captain Farragut was the onboard commander of the USS *Hartford* and the commander of all the 24 wooden-hulled ships in his fleet. Farragut's fleet passed the forts after a noisy and bloody day and night battle, as other naval and military forces remained in the immediate proximity to continue to subdue the formidable forts and the courageous Confederate gunners. On 25 April 1862 New Orleans surrendered and was occupied by U.S. Army troops and U.S. Marines as U.S. Navy vessels stood guard on the Mississippi River approaches to the South's largest port city. For Farragut's success, the U.S. Congress commissioned him to the rank of rear admiral.

In 1862 Farragut continued his patrol of the Mississippi River and participated in an unsuccessful U.S. Navy and U.S. Army attack on the port of Vicksburg, Mississippi. Admiral Farragut then sailed south on the Mississippi River into the Gulf of Mexico to attack the Confederate forts and naval vessels at Mobile Bay, Alabama. The attack commenced on 5 August 1864. The heavily

The 73-foot RC *Morris* carried six 9-pound guns and a crew complement of twenty. Cuttermen prepare to board the passenger ship *Benjamin Adams* 200 miles east of New York on 16 July 1861.

torpedoed (mined) Mobile Bay claimed the USS *Tecumseh*. After that sinking the lead ships of the fleet paused. Observing the event, the courageous admiral climbed aloft into the rigging of his flagship, the USS *Hartford*, and motivated the hesitant sailors to continue the attack, with his famed shouted order, "Damn the torpedoes! Full steam ahead!"[12] Farragut's squadron confronted the notorious Confederate commander, Admiral Franklin Buchanan (CSN), his three gunboats, and the ironclad CSS *Tennessee*.

Buchanan had served in the U.S. Navy as the first superintendent of the U.S. Naval Academy (1845–1847), fought in the Mexican-American War, was commandant of the Washington Navy Yard, and commanded the USS *Susquehanna*. Upon joining the CSN in 1861 then Captain Buchanan commanded the iron-hulled monitor CSS *Virginia* (former USS *Merrimack*) on 8 March 1862 in the battle at Hampton Roads, Virginia, where the *Virginia* defeated the wooden ships USS *Congress* and USS *Cumberland*. Wounded in that battle, Capt. Buchanan did not command the *Virginia* the following day in the standoff with the USS *Monitor*. In 1862 Admiral Buchanan was put in command of the forces at Mobile Bay and assigned to the ram *Tennessee*. After the Civil War he assumed the presidency of the Maryland Agricultural College (1866) in his home state. He passed away in 1874.[13]

In 1829 David Dixon Porter joined the Navy. By April 1861 Cmdr. Porter was in command of the USS *Powhatan*, which patrolled the Gulf of Mexico. In 1862 he led a mortar-boat flotilla under Capt. Farragut's command in the battle against the Confederate Forts Jackson and St. Philip. In October 1862 Porter assumed command of the Mississippi Squadron, which would finally capture Vicksburg (1863) in a joint operation with the U.S. Army. The victory led to his promotion, and now Rear Admiral Porter was assigned command of the North Atlantic Blockading Squadron. At the end of the Civil War Porter assumed new responsibilities as superintendent of the U.S. Naval Academy and then was appointed the head of the U.S. Navy Department, earning the rank of full admiral in 1870. Admiral Porter crossed the final bar in 1870.[14]

We now return to the naval battles at Fort Jackson and Fort St. Philip on the Mississippi River in Louisiana (April 18–28, 1862). David D. Porter had been in command of a mortar gunboat flotilla in Captain David Farragut's armada that moved up river between the forts to capture New Orleans. The forts, one on each bank of the river, possessed heavy artillery batteries and were solidly built with masonry and timber and earthen ridges. A line of sunken hulks and an iron chain blocked the passage of ships. Ten Confederate naval vessels served as the defense fleet of the two forts. Two of the warships (CSS *Louisiana* and CSS *Manassas*) were heavily armed ironclads, but inadequate engine power forced the CSS *Louisiana* to moor and anchor against the shore.

Captain Farragut (USN) had 24 wooden-hull warships under his command. Commander Porter directed 18 mortar schooners, each ship mounting a devastating 13-inch mortar cannon that could shoot a 200-pound shell more than two miles, depending on the angle at which the weapon was set. Farragut's mobile squadron and dark-of-night maneuvers made difficult targets for the Rebel gunners. The Federals managed to cut the chain and move the hulks enough to allow the fleet to move past the forts but not without ship damage and casualties and confrontations with the formidable ram CSS *Manassas* and floating fire rafts. One of the current-powered rafts ignited a conflagration that almost sank Farragut's ship, the USS *Hartford*, and would have but for the fire-fighting courage and skills of quick-acting enlisted sailors.

From the forts Captain Farragut sailed 70 miles upstream and went ashore on April 25 to supervise the surrender of New Orleans after confronting hostile crowds and belligerent officials. On the 28th the commander of Forts Jackson and St. Philip, Brigadier General Johnson K. Duncan (CSA), surrendered.[15] On 28 April Gen. Duncan sent a Confederate officer by boat to the Revenue

Cutter *Harriet Lane*, which served as Cmdr. Porter's flagship. Professor Benton Rain Patterson has traced the Mississippi River campaign between 1861 and 1863 and states in his book that the surrender documents were signed onboard the *Harriet Lane* on 29 April 1862, the proceedings being interrupted by a situation that could have been disastrous for the crew members of the revenue cutter and the Union and Confederate officers who were to sign the articles of surrender, Commander Porter and General Duncan. The huge Confederate ironclad *Louisiana*, adrift and in flames, exploded, killing a soldier at Fort St. Philip and doing additional destruction to the bastion and causing the *Harriet Lane* to roll in the wake of the blast.[16] Fort Jackson casualties totaled 33 wounded and 9 killed. The casualty count at Ft. St. Philip was 4 wounded in action, and 2 military personnel killed. The Confederate naval personnel casualty count is uncertain, but one source calculated 73 CSN sailors were killed in action and an equal number were wounded. One Confederate warship itself calculated that 57 of its crew were killed. United States Navy records indicate a casualty count of 147 wounded and 37 were killed in action.[17]

John Dahlgren was a naval ordnance expert and innovator and the commander of the South Atlantic Blockading Squadron during the War Between the States. He joined the U.S. Navy in 1826 and spent several years working with the U.S. Coast Survey, as he did the U.S. Revenue Marine. The two naval branches (USN and USRM) did supportive work with the U.S. Lighthouse Service, and U.S. Life-Saving Service. In 1847, Lt. Dahlgren assumed ordnance duties at the U.S. Navy Yard in Washington, D.C., where he invented boat guns and smooth-bore and rifled (grooved barrel) warship deck guns. The rifled barrels give a spiral spin to the projectile or shot, which allows greater target accuracy. In 1855 Dahlgren was appointed as a commander, a captain in 1862, and a rear admiral in 1863. In 1861 and 1862 Rear Adm. Dahlgren was the commandant of the Washington Navy Yard and the Bureau of Ordnance chief. In 1863 he took command of the South Atlantic Blockading Squadron and spent two years waging naval warfare and blockade missions in proximity to the port of Charleston, South Carolina. When the Civil War ended Admiral Dahlgren's assignments included commanding the South Pacific Squadron, Washington Navy Yard, and the Bureau of Ordnance. His ordnance innovations contributed a significant advantage to the Union over Confederate military and naval power. Admiral Dahlgren passed away just five years after the Civil War ended.[18]

When the Southern states seceded from the Union to form the Confederate States of America they had a minimal-strength navy, but 236 U.S. Navy officers out of a total of 1,457 resigned their commissions to join the Confederate States Navy (CSN). The U.S. Navy Yard at Norfolk, Virginia, was evacuated by American naval personnel and seized by Confederate forces on 20 April

1861. The Federals destroyed as many vessels and as much infrastructure as possible in the time they had, but the Confederates acquired invaluable naval supplies, ordnance, and ship remnants. The USS *Merrimack* was raised and restored and renamed the CSS *Virginia*. The 263-foot–long ship was outfitted with a cast-iron ramming plow on the bow stem and, as mentioned earlier, was victorious over U.S. naval ships at Hampton Roads, Virginia, on March 8–9, 1861. The iron-hulled USS *Monitor* subsequently battled the CSS *Virginia* to a draw, and the U.S. fleet saved the grounded USS *Minnesota* from destruction and sinking by the enemy. The *Virginia* skipper, Commodore Franklin Buchanan (CSN), was wounded[19] but survived to fight and be defeated by Admiral David Farragut (USN) at Mobile Bay, Alabama, in August of 1864.[20]

Southern blockade-runners hid out in secluded coves and estuaries and made successful runs past Union blockading vessels. Blockade-runners are estimated to have brought $200 million worth of imports into the CSA, and they exported more than $1 million in various commodities and agricultural crops, including cotton. By 1864 the U.S. Navy had stopped most Southern maritime commerce, and the blockade-runners and their hundreds of fast vessels continued facing "the dangers of blockade running for the sake of patriotism [and] profit."[21] Confederate blockade-running ships generally sailed to and from such Western Hemisphere insular British colonies as Bermuda and the Bahamas and the Spanish colonies of Cuba and the Dominican Republic, bringing home luxury goods, ammunition, guns and medicines.[22]

In the Civil War, the naval blockade significantly separated the CSA from Europe commercially and diplomatically. Dominance at sea and on coastal and internal waterways by the U.S. Navy and U.S. Revenue Cutter Service allowed U.S. Army, militia and volunteer units to move great distances by ocean and river transport (steam-powered) vessels, and join the U.S. Navy and U.S. Marines in amphibious (ship- and boat-to-shore) assaults. Steam- and even sail (wind)-powered Union warships sailing in circles against coastal forts and drifting past river fortifications proved that shore-based bastions under Confederate army control could be defeated with minimal harm to Union vessels, although naval ships, boats, and crews did suffer significant damage and casualties.

In January 1865 forty-four U.S. Navy warships bombarded Fort Fisher in North Carolina at the mouth of the Cape Fear River. The conquest of Confederate forts and ports deprived the South of their use and provided coaling stations for the steam-powered Union fleet. Cape Hatteras Inlet (North Carolina) was one of the earliest Confederate forts seized and occupied by the U.S. Navy. In 1861 and 1862, USN warships defeated and occupied the Confederate ports and forts at Hatteras Inlet and Roanoke Island off Virginia's coast.[23] Port Royal Sound housed Fort Beauregard on St. Helena Island and Fort Walker on Hilton

Head Island off South Carolina. The CSN defensive force consisted of seven small armed steamers, mostly tugboats, under the command of Commodore Josiah Tattnall (CSN), who had to face Flag Officer Samuel F. Du Pont (USN) and his 75-vessel squadron, which included steam frigates, one dozen smaller steam vessels, 50 transports and colliers (coal carrying ships), and a few sailing warships.

The Battle of Port Royal commenced and terminated on 7 November 1861, after the outnumbered naval and military confederate personnel surrendered after putting up the best fight they could with the ordnance, ships and boats, and men they had. Among the USN vessels that contributed to the victory were the USS *Wabash, Susquehanna,* and *Pocahontas,* the latter skippered by Cmdr. Percival Drayton (USN), whose brother (Brig. Gen. Thomas E. Drayton, CSA) was the commanding officer of Fort Walker. The shore-side regional Confederate commander was General Robert E. Lee (CSA).[24]

The Mississippi River war at Vicksburg, Mississippi (December 1862– July 1863), was a coordinated U.S. Navy–U.S. Army victory of strategic significance. Vicksburg and its protective array of artillery batteries ran for several miles at a bend in the river along a 200-foot–high bluff. The bluff ran north and south at Vicksburg. But the Mississippi River bends north and south of the city and the rest of the surrounding physical geography made an assault on the city from the river nearly impossible. Admiral Farragut had sailed past Vicksburg after his New Orleans victory in 1862, but that naval presence did not intimidate the Mississippi city into surrendering. Confederate engineers increased the invulnerability of the city by adding to the gun emplacements between 1862 and 1863. Nonetheless, Vicksburg would be conquered eventually because of the leadership skills and cooperation between Rear Admiral David Porter (USN), the Mississippi Squadron commander, and then Major General Ulysses S. Grant (U.S. Army).

On 16 April Admiral Porter consented to General Grant's problematic request for Porter to test the strengthened Confederate batteries by sailing past them in the dark of night. Alerted Confederates descended the river in boats to light bonfires on the far shore of the Mississippi River in an attempt to show the American warships as silhouettes along the riverbanks. Confederate artillery fire from the high bluffs caused the loss of several barges and one transport ship. Rear Adm. Porter's boats then maneuvered to land Gen. Grant's troops onshore a few miles south of Vicksburg. General Grant's troops then marched inland to capture strategic territory, including the state capital of Jackson on May 14, 1863. Grant's army then turned west back toward the Mississippi River, fought a seven-week battle at Vicksburg on the bluffs, and accepted the surrender of the Confederate Army at that location on the 4th of July 1863.

The joint mission of the U.S. Navy and U.S. Army completed the Union quest for control of the Mississippi River and the fall of Vicksburg.[25]

The Battle of Charleston, South Carolina, on 7 April 1863 was significant to Union and Confederate military and naval forces and political leaders because South Carolina was the first state to secede from the United States, and the Civil War commenced with the firing by the Confederates on Fort Sumter in Charleston Harbor. Charleston was also the hub of the blockade-running. Union strategists therefore were determined to finally capture Charleston, an objective that had eluded them since the outbreak of the Civil War.

U.S. Navy secretary Welles thought Rear Admiral Du Pont (USN) and his South Atlantic Blockading Squadron, given its recent acquisition of several new ironclad warships, should finally be able to sail past Fort Sumter, take control of Charleston Harbor, and suppress the Confederate blockade-running ships. But Rear Adm. Du Pont was reluctant to carry out that mission because, since 1861, the CSA had added hundreds of heavy guns to the harbor arsenal, saturated the harbor waters with mines, and positioned an enhanced ironclad fleet in strategic locations. The strengthened Confederate strategic and tactical position at Charleston was illustrated by a 31 January 1863 CSN attack from two ironclads that badly damaged two blockading Union warships.[26]

Finally, on April 6–7, 1863, a reluctant Admiral Du Pont sailed his fleet into Charleston Harbor behind the lead vessel USS *Weehawken*, which had been outfitted with an anti-mine raft off the bow equipped with hooking devices ostensibly positioned to disable and reposition the mines in order to render the formidable ordnance ineffective. Unfortunately, the *Weehawken's* anchor components got hung up with the grappling hooks on the raft, and the channel currents had moved Rear Adm. Du Pont's flagship, the USS *New Ironsides*, into shoal waters, forcing him to anchor right over an enemy torpedo mine, its electric detonator fortuitously failing to ignite the torpedo. The other ships of the Union fleet advanced into Charleston Harbor firing on, and being fired at from, the several area forts and shore gun emplacements during the two-hour battle. An estimated 200 shots were fired at Confederate positions on 7 April, with about 25 percent of the projectiles hitting targets. The U.S. vessels endured heavy fire from the forts. The USS *Patapisco*, *Weehawken*, and *Nantucket* were each hit by an average of around 50 enemy shots. The USS *Keokuk*, an ironclad with two gun towers, sank after enduring a barrage of 90 Confederate hits.

The battle damage reports caused Admiral Du Pont to call off a renewed attack scheduled for the following day. When word reached Secretary Welles that the attack was terminated he replaced Admiral Du Pont with Rear Admiral John A. Dahlgren the following month. But Adm. Dahlgren agreed with Du

Pont and did not renew a frontal attack on the harbor forts. Instead, Dahlgren joined the troops of Major General Quincy A. Gilmore (U.S. Army) in a joint attack against Fort Wagner and what was called Battery Gregg on the South Carolina coast immediately south and outside of Charleston Harbor. On 6 September 1863 Fort Wagner surrendered after Union troops suffered heavy casualties. U.S. Army troops were then able to fire upon Fort Sumter from Battery Gregg for the remainder of the year. But even then, Charleston did not surrender until nearly the end of the war when it was assaulted from land by the troops of General William T. Sherman (U.S. Army) during his "March to the Sea."[27]

11

The Revenue Cutter Service
in the Civil War

In April of 1861, just before the Civil War commenced, the U.S. Revenue Marine, later the Revenue Cutter Service, had only 28 cutters ready for wartime missions. Four of these cutters were sailing off the Pacific Coast; 18 were stationed out of Atlantic Coast ports; and six were patrolling the "Inland Seas" of the Great Lakes. During the buildup and then outbreak of the Civil War, Revenue Marine officers sympathetic to the South turned five cutters over to Confederate authorities who used them in the Confederate States Navy (CSN) as combat vessels, blockade-runners, and navy or private commerce raiders.

President Abraham Lincoln ordered Treasury Secretary Salmon B. Chase to use the cutters to assist the United States Navy, as the Revenue Cutter Service had done since 1798. Secretary Chase ordered customs collectors and revenue cutter officers to cut off or interdict maritime commerce into and out of the Confederate States of America. Great Lakes cutters were ordered to sail to Atlantic Coast ports for subsequent wartime duties. Cutters given previously to the U.S. Coast Survey were returned to the USRCS. Other boats and cutters were constructed and paid for by the federal government, and donated by private citizens. Besides assuming combat duties, the cutters continued their traditional duties of collecting revenues, carrying out search and rescue missions, and protecting cargo and fishing vessels that were targets of rebel commerce raiders, including the RC *Caleb Cushing*, which was sunk by a Confederate warship off of Portland, Maine, on 27 June 1863.

The Revenue Cutter Service had one side-wheel steam vessel, The RC *Harriet Lane*, otherwise depending on wind-powered sailing ships. Sent to resupply Charleston Harbor, the *Harriet Lane* fired the first naval shots of the Civil War, off Ft. Sumter, South Carolina, and later engaged in battle off Cape Hatteras and on the Mississippi River. In January 1863 the cutter was captured by Confederate naval and military forces at the port of Galveston, Texas, became the CSS *Harriet Lane*, and then served as a rebel blockade-running vessel.[1] The *Harriet Lane* is identified in history as the USRC, USS, and CSS *Harriet Lane* because it served as a revenue cutter under the Revenue Cutter Service, was transferred to the jurisdiction of the U.S. Navy as a U.S. Navy Ship, and then was captured and used by the Confederate States Navy as a Confederate ship. Built in 1857 and launched in New York, the side-wheel steamer was named for President James Buchanan's niece. The coal-fired steamer carried enough fuel to last about a week, but its sail superstructure allowed the auxiliary vessel to use wind power at sea and thus ration its coal supply. The *Harriet Lane* fired that first naval shot of the war on 12 April 1861 across the bow of the *Nashville* when that ship ignored the *Harriet Lane*'s signal to come to and show her colors. For several weeks the *Harriet Lane* remained in the immediate region serving with the U.S. Navy as a blockade ship stationed off Fort Monroe, Virginia. The *Harriet Lane* was part of the U.S. Navy fleet on the lower Mississippi River that subdued the Confederate Forts Jackson and St. Philip (April 18–28, 1862) in the successful Union campaign to capture the city of New Orleans and split the South into two regions.[2]

The *Harriet Lane* and five steam-powered riverboats towed the schooner mortar boats that bombarded Forts Jackson and St. Philip during the naval assault that preceded the upriver mission to New Orleans. The assault on the forts was terminated when Brig. Gen. Johnson K. Duncan (CSA) surrendered to Commander David Dixon Porter (USN) on the *Harriet Lane*, which was Porter's flagship.[3] A formidable vessel with a distinguished and complex history, it carried a crew complement of 95; could do 13 knots; displaced 600 tons; was 177 feet in length; and was powered by two masts and two side-wheel paddles. The cutter's armament included 4 guns ranging from 4 to 8 to 9 inches and two brass 24-pound howitzers. The hull was sheathed with copper. In January 1863 the *Harriet Lane* was captured by Confederate naval and military forces in the port of Galveston and used as a blockade-runner against the U.S. Navy's blockade of Southern ports. In 1864 the now Rebel blockade-runner made it to Spanish Cuba with a load of contraband cotton. In 1867 Spanish officials returned the *Harriet Lane* to United States authorities.[4]

In his classic *Definitive History of the United States Coast Guard (1790–1915)*, published in 1949, historian and Captain Stephen H. Evans (USCG) credited U.S. Navy designers and board members William H. Webb, Charles W. Copeland, Henry Hunt, Francis Grice, and Samuel M. Pook with the naval architectural ideas and skills that made the side-wheel and sail-steamer USRC *Harriet Lane* (1857) unique and superbly suited for its subsequent missions. The 674-ton, 180-foot cutter possessed the "fast lines" of advanced sailing ships

Another view of the vaunted paddle-wheel, steam-powered, wooden-hulled USRC *Harriet Lane* of Civil War fame. The *Harriet Lane* would be turned over to U.S. Navy control and later be captured by Confederate forces.

with its 12-foot depth, 10-foot draft, 30-foot beam, and 12-knot speed. Captain John Faunce (USRM), who was the first captain of the *Harriet Lane* and was in command of the cutter on 12 April 1861 in Charleston Harbor, reported his appreciation of that "noble vessel" to the builder, Mr. Webb, and the board. Its first mission was with the U.S. Navy under Commodore William B. Shubrick (USN), the commander of the 15-vessel fleet that completed the Latin Americas mission in 1858. The *Harriet Lane* was in combat action with the U.S. Navy, U.S. Marines, and U.S. Revenue Marine. More than 1,300 sailors manned the fleet. The *Harriet Lane* carried around 100 men and 7 guns.[5]

The cutters, despite an initially small Treasury Department fleet of 10 to 12 ships, were busy enforcing the essential revenue laws that provided the federal government with the resources to finance the war, as well as performing military support with the U.S. Navy to secure the eventual Union victory over the Confederacy. Captain H.B. Nones (USRM) commanded the old cutter *Forward*, a Mexican War vessel, on Chesapeake Bay with orders to attack enemy vessels. The cutter fleet was expanded with congressional appropriations to construct new ships or purchase vessels from private individuals and companies. Among the fortuitous purchases was the donation of the RC *Naugatuck*, also known as the *E.A. Stevens,* from Mr. Stevens. The vessel was a 100-foot, twin-screw, steam-powered, semisubmersible cutter that saw naval combat action. The Revenue Marine acquired six other single-screw steamers in 1863.

The RC *Miami,* formerly a civilian yacht, steamed down the Potomac River in May 1862 carrying Federal notables General E.L. Viele of the U.S. Army, Secretary of War Edwin M. Stanton, Treasury Secretary Salmon P. Chase, and President Abraham Lincoln. The secret mission placed the Federal entourage on shore at Norfolk, Virginia, to ascertain why a Union attack on Confederate territory had not commenced. The U.S. Army generals had stalled, thinking an amphibious attack with soldiers, marines, and army personnel could not be launched because of shoal waters and sandbars. Lincoln's presence onshore ended that debate. The Commander in chief ordered a covering naval bombardment and transport attack the next day that was successful. On 15 May 1862, a USN-USRM squadron pushed on up the James River to challenge the

enemy at the Confederate capital of Richmond, Virginia, and to assist Union troops. The conquest of Richmond would occur at the end of the war in 1865, but the cutters *Naugatuck* and *Harriet Lane* later escorted hundreds of troops who battled Confederate forces with courage, skill, and good results. Captain Faunce would write a detailed report, dated 5 June 1861, to Flag Officer G.J. Pendergrass (Squadron Commander, USN) detailing the successful attack on Fort Monroe. The naval battle off Hatteras Inlet on August 28–29, 1861, resulted in the capture of Confederate forts, enhanced Union morale

Captain H.B. Nones (USRCS) was born in Virginia in 1804 and entered the USRCS as a 2nd lieutenant in 1831. Known for his many dangerous rescues, Nones was also noted for his long command of the Revenue Cutter *Forward,* from 1841 through the Mexican War (1846–1848) and the Civil War (1861–1865). During the Mexican War Captain Nones led naval attacks on two ports in Mexico and played a major role in the blockade of the port of Vera Cruz. He died in 1868.

Captain Henry Benjamin Nones of the U.S. Revenue Cutter Service in the full dress uniform of a lieutenant in this Civil War era (Library of Congress).

after the defeat suffered by Union forces at Bull Run in July 1861, and convinced several European nations not to provide the CSA with significant foreign support.

Flag Officer Silas Stringham (USN) combined his naval forces with U.S. Army units, the U.S. Marine Corps, and U.S. Revenue Marine. Two purchased transports (SS *Adelaide* and SS *George Peabody*) carried one thousand U.S. Army troops under the command of Major General Benjamin Butler. The Navy and Revenue Marine vessels included the USS *Minnesota, Wabash, Pawnee, Monticello, Cumberland,* and *Susquehanna,* and the RC *Harriet Lane,* now under USN jurisdiction and properly identified as the USS *Harriet Lane,* under the able command of Captain John Faunce (USRM). The rifled guns of the *Harriet Lane* hit the Confederate Battery and the forts. After a valiant defensive response, several Confederate officers (Commodore Samuel Barron, CSN, Col. William F. Martin, CSA, and Major W.S.G. Andrews, CSA) signed articles of capitulation on Commodore Stringham's flagship. The U.S. Army commander, Maj. Gen. Butler, also signed the surrender document.

The U.S. Lighthouse Establishment, including ship personnel, lighthouse keepers, and crews, supported the Revenue Marine, contributed to successful Civil War missions by repairing lighthouses captured or destroyed by Rebel forces, and recaptured or replaced buoys, lights, and lightships in the maritime coastal and inland waters of the theaters of operations. U.S. lightships were armed with rifled deck guns and nets strategically placed above decks to repel Rebel boarding attempts.[6]

Dr. Robert Browning, Jr., then the chief historian of the U.S. Coast Guard at the Washington, D.C. headquarters, wrote a magnificent history of the Union South Blockading Squadron in the Civil War. The USN Squadron stretched from the south Atlantic coastline to around Florida. The naval squadron was tasked with engaging Confederate commerce raiders and warships, and supporting amphibious assaults against Southern forts, enemy ironclad ships, and the CSN submarine, CSS *Hunley.* As with his previous book about the North Atlantic Blockading Squadron, the historian chronicles battles, tactics, CSN and USN strategy and leadership, shipbuilding and logistics. In his discussion of the Revenue Marine's contributions to U.S. Navy operations, Browning described how the Union supply ship *Star of the West* (Captain John McGowan, a former Revenue Marine officer who rejoined the USRM shortly after that unsuccessful supply run), crept into Charleston Harbor on 9 January 1861 with 200 troops onboard to support and provision the besieged Ft. Sumter, which was under the command of Major Robert Anderson. Captain McGowan advanced against Confederate gunfire, concluded the two-mile harbor journey to Ft. Sumter would be impossible to survive, and headed back out to sea.

Two months later, President Abraham Lincoln, President Buchanan's successor, informed Confederate political and military authorities that he would simply try to replenish the dwindling supplies of Ft. Sumter for the U.S. Army commander, Major Robert Anderson. On 12 April 1861, Anderson, who refused a CSA demand to surrender, endured Rebel shelling. The USS *Harriet Lane* (Captain John Faunce, USRM) was forced to withdraw from Charleston Harbor after firing the first naval shot of the Civil War. After putting up a gallant defense and expending its ordnance, Major Anderson surrendered to the CSA. The *Harriet Lane*, accompanied by the supply ship SS *Baltic*, carried the defeated Union troops back to Washington, D.C. Ft. Sumter, despite numerous naval bombardments, would not fall back to Union forces until near the end of the Civil War in 1865.[7] The South Atlantic Blockading Squadron belatedly prevailed late in the war "because," Browning concluded, "the leaders in Washington failed to clearly express their strategic vision. The record of military actions along the South Atlantic Coast [was] full of bravery, sacrifice, and victory, as well as unfulfilled expectations, lost opportunities, and failure." But in the end, as Browning titled his book, "success is all that was expected."[8]

Flag Officer Garrett J. Pendergrast (USN), a veteran of the War of 1812, was assigned the responsibility of commanding the Atlantic blockade, which spanned 900 miles from Chesapeake Bay to Florida waters. The USS *Cumberland*, Pendergrast's flagship, was stationed off Fort Monroe. The revenue cutter *Harriet Lane* was one of the ships in the Atlantic Blockading Squadron in the maritime area of Charleston, South Carolina, along with other steamers that included USS *Niagara*, *Seminole* and *Wabash* and the sloop USS *Vandalia*. The steamships had to sail back and forth from Pensacola, Florida, and Hampton Roads, Virginia, to replenish their coal supplies that were burned up on the blockade patrols off Charleston Harbor. On occasion, for example on 19 May 1861, the RC *Harriet Lane* would be the lone naval vessel off Charleston, until the cutter was joined by the 264-foot, 23-foot draft, screw frigate USS *Minnesota*, the flagship of Flag Officer Silas Stringham, the commander of the Atlantic Coast Blockading Squadron. The Blockade Squadron vessels varied in time on station due to differences in coal-carrying capacity and the time it took to sail to coal yards and back, the differential drafts of the vessels that determined how close to shore and shoal waters the ships could get, and how far out to sea they were stationed.[9]

Dr. Browning completed a follow-up history of the West Gulf Blockading Squadron in the Civil War, entitled *Lincoln's Trident*,[10] a detailed and scholarly history that has been appropriately described as "magisterial." Established by the U.S. Navy Department in 1862, the West Gulf Blockading Squadron patrolled the expansive Gulf of Mexico maritime region that stretched from the coast of West Florida to the Rio Grande River. Browning described the devastating impact of the U.S. Navy and U.S. Revenue Cutter Service on Southern commerce, troops, supplies, and CSN vessels in the Gulf, lower Mississippi River, and the port of New Orleans, Louisiana.

Union navy commanders and crews and U.S. Army troops cooperated in successful operations from the Gulf, up the Mississippi to New Orleans and Vicksburg, and into the CSN and CSA strongholds around Mobile Bay, Alabama. The USRC/USS *Harriet Lane* performed valiantly in the theater of operations of the U.S. Navy Squadron commanded by Admiral David Farragut (USN) on his formidable screw/sail sloop and flagship, USS *Hartford*. In the attack on the Confederate forts on the Mississippi on the way up river to New Orleans, Adm. Farragut boarded the RC *Harriet Lane* in a reconnaissance trip to inspect the obstruction boom the Rebels had built between Forts Jackson and St. Philip.[11] Following the capture of New Orleans, the *Harriet Lane* would accompany Adm. Farragut to Vicksburg, Mississippi, in 1862, but victory there would have to wait for a future assault in 1863. Robert Browning wrote that the Navy Department's West Gulf Blockading Squadron had to patrol a coastline extending more than one thousand miles until Union naval forces defeated the naval and military forces of the Confederate States of America at Mobile Bay in 1864.

Browning's creative description of the Union blockading squadrons on the Atlantic and Gulf coasts and on the lower Mississippi as "Lincoln's Trident" makes reference to the classical story of the Greek god Poseidon, "who struck the ground with his trident to cause earthquakes, tidal waves, and storms at sea."[12] President Lincoln's "Trident" consisted of the Union naval and military forces that carried out General Winfield Scott's "Anaconda Plan" to crush the South by using naval and ground forces to geographically divide and crush the war-making capacity of the Confederacy on land and sea through absolute Federal control of strategic land and maritime regions.[13] The chief historian of the Coast Guard concluded that the U.S. naval and military forces of the Blockading Squadron shortened the time frame of the Civil War and were significant factors in the achievement of the Union victory over the Confederate States of America[14] and the reunification of the United States of America.

Despite that ringing endorsement of the strategy and impact of the Blockade Squadron mission along the coasts and rivers of America in the Civil War, Browning objectively assessed its significant deficiencies. Federal, naval, and military administrative failures initially hampered the blockade efforts. Among the failures was the initial lack of mission coordination and strategies between soldiers and sailors in interior and coastal waters. Experience would later ameliorate that situation. President Lincoln did improve the situation with effective and perceptive vision in communications with his navy and army secretaries and leaders and his ability to effect compromises between the different strategic and tactical visions of the respective armed services. The U.S. Army needed the U.S. Navy to transport its forces within the convoluted and differing environments of the vast interior and coastal regions. The U.S. Navy and U.S. Revenue Cutter Service required the time, planning, and resources necessary to navigate the blue waters and the brown waters of the American maritime regions. Personnel had to be trained to master the plethora of skills needed to operate the ships, boats and weapons and to meet the challenges of naval warfare in a variety of geographic situations and meteorological variables. Military and naval personnel had to master combat skills and be effective once the horror, danger, blood and carnage of war commenced in each battle. Present as well were the inevitable conflicts that ensued because of interservice rivalries; competition for resources, rewards and recognition; and the ego and personality conflicts between naval and military leaders and political leaders.

The Union forces had a distinct advantage in the vast domain their offensive operations operated in. The Confederate forces, while enjoying certain defensive advantages, suffered from their inability to distribute supplies and ordnance throughout their maritime and terrestrial domains because of limited railroad connections between production sites and coastal and river ports and the disadvantage the South had in competing with the overwhelming infrastructure and industrial production and population dominance of the North. Nonetheless, the South was able to export and import necessary supplies to a surprising extent, given the vast geographic expanse the Union blockade had to cover. Their successes caused some observers to refer to the blockade as a "sieve," facilitated by the tactical advantage Rebel captains had with their knowledge of deep and shoal waters and the maritime routes their ships commanded between American and West Indies and European ports.[15] But overall, Browning concluded, "The West Gulf Blockading Squadron did a creditable job ... given the number and types of warships available ... and ... succeeded in adding to the South's isolation, and kept the Confederacy from establishing a full-scale war economy."[16]

On the issue of revenue cutter officers joining the Confederate States of America, examples are significant. Secessionist fever was in the air before the commencement of the Civil War with the firing on Ft. Sumter. Captain James J. Morrison (USRM) served in the Seminole Wars and the Mexican War. Captain Morrison transferred the RC *Lewis Cass* out of Mobile to pro–Confederate officials in that state, forcing crew members loyal to the United States to trek, under the leadership of Third Lieutenant Charles F. Shoemaker, across the South into Union territory. Lt. Shoemaker eventually became the captain-commandant of the U.S. Revenue Cutter Service.

The RC *McClelland* operated out of New Orleans under the command of Captain John G. Brushwood. The Louisiana Revenue Marine officer was aware of the Southern sympathies that prompted U.S. Treasury secretary John A. Dix to order the cutter commander to proceed with his ship and crew to the port of New York City. Upon Brushwood's refusal to follow those orders, Secretary Dix wired the officer who was second in command on the *McClelland* to take Capt. Brushwood into custody, seize command of the cutter, and sail to New York City, with further orders to shoot any sailor who tried to lower the U.S. flag on the vessel. The second officer in command refused the orders and assisted Capt. Brushwood in lowering the Stars and Stripes and raising the CSA flag aloft. A Revenue Marine cutter was destroyed during the Rebel take-over of the Norfolk (Virginia) Navy Yard, and three more cutters fell to Confederate sailors in ports throughout the South.[17]

In 1863 the legendary USRC *Harriet Lane* was lost to the Revenue Marine and U.S. Navy when it was attacked and overrun by Confederate military and naval forces while tied up at the pier in the port of Galveston. The veteran commander Capt. John Faunce (USRM) was not on board but had been called to New York to examine candidates for commissioned officer rank. The U.S. Navy commander on board and his crew were heavily outnumbered and outgunned but fought valiantly. Union sailors were killed in the battle as two Confederate steamships smashed into the vulnerable side-wheels of the cutter and Rebel soldiers boarded the vessel and fought hand to hand on deck. The *Harriet Lane* flew the flag of the CSA for the remainder of the Civil War.[18]

The wartime shoreline and sea combat, revenue collection, and ship and crew rescue missions of the U.S. Revenue Cutter Service forced Congress to appropriate the necessary funds for the construction of six new, 135-foot steam-propelled, single-screw, six-gun cutters. The positive notoriety the USRCS gained through wartime publicity prompted a prescient writer in the *Army and Navy Journal* (1864) to predict the future Coast Guard's name and motto ("Always Ready," or Semper Paratus): "Keeping always under steam and ever ready in the event of extraordinary need to render valuable service, the cutters can be made to form a coast guard whose value it is impossible at the present time to estimate."[19]

Among the many challenging and interesting missions assigned to the USRCS during he Civil War was the order to search all outbound vessels to apprehend President Lincoln's assassins and sympathizers after his assassination on 15 April 1865.[20] The Coast Guard historian Truman Strobridge reviewed several of the Civil War events involving the USRCS that we have previously considered. Upon the entry of the federally charted merchant ship SS *Star of the West*, under the command of former (and future) USRM Captain John

McGowan, into Charleston Harbor and the first naval skirmish between the Federals and Confederates, Strobridge quotes McGowan as recounting that a Rebel cannonball "ricocheted off the water and over" the unarmed side-wheel ship: "You must give us bigger guns that that, boys, or you'll never hurt us."[21] Then a Confederate cannonball hit the hull of the freighter, forcing McGowan to reluctantly withdraw from the harbor. Captain McGowan was unarmed but knew combat from his previous Revenue Marine career fighting smugglers and pirates and interdicting slave ships. He had hoped to bring his reinforcement contingent of U.S. Army soldiers and supplies to Ft. Sumter and gain access as a coastal trade vessel.[22]

Regarding the first naval shots fired at Fort Sumter by the USRC/USS *Harriet Lane*, Strobridge cited the writing of G.S. Osbon, a Civil War correspondent who was onboard the cutter: "A vessel suddenly appeared through the mist ... a passenger steamer made out to be the (SS) *Nashville*. She had no colors [flag] set, and as she approached the fleet she refused to show them. Captain Faunce ordered one of the guns manned, and as she came still nearer, [Faunce] turned to the gunner. "Stop her," he said, and a shot went skipping across her bow. Immediately, the U.S. ensign went to her gaff end and she was allowed to proceed. The *Harriet Lane* had fired the first gun from the Union side.[23] Thus the U.S. Revenue Marine is credited for firing the first naval (and Federal) shot in the Civil War. The Revenue Marine had been known since the time of its first civilian leader, Treasury Secretary Alexander Hamilton, by several service names, including "A System of Cutters," "Revenue Marine," Revenue Marine Service," and "Revenue Cutter Service." Strobridge informs us, "The title Revenue Cutter Service found its first statutory use in an Act passed in 1863."[24]

12

Historical Assessments
of the Civil War Navies

Numerous historians have applied their historiographical and analytical skills to the complexities of missions and leadership of the Union and the Confederacy navies. In Chapter 11, we reviewed the scholarship of Coast Guard historians Truman Strobridge and Dr. Robert Browning, Jr. Strobridge traced the actions of specific revenue cutters. Browning chronicled the tactical and strategic operations of the U.S. Navy Blockading Squadrons in the Atlantic and Gulf regions. In this chapter, we shall consider the perspectives of other historians about events in this complex period of history. Historians continually revisit the Civil War in new books and periodicals as different evidence, insights and documents emerge. The process of historical revisionism is ongoing because interpretations change with the perspectives of time and new generations.

This author and coauthor John J. Galluzzo studied several of the commissioned officers who became United States Revenue Marine/Revenue Cutter Service and Coast Guard leaders and their missions and strategies in the Civil War.[1] In the foreword to the Ostrom/Galluzzo book *United States Coast Guard Leaders and Missions, 1790 to the Present*, Captain W. Russ Webster, USCG (Ret.) assessed the leadership and mission contributions of the USRCS and USCG in domestic and overseas actions. Webster described the historic Revenue Cutter and Coast Guard "interconnectedness with the Navy and presidents" of the United States.[2] John Galluzzo described how "Captain John Faunce (USRCS), on the cutter *Harriet Lane*, fired the first naval shots of the Civil War outside Charleston Harbor" in April of 1861.[3] This author chronicled the revenue cutter voyage down Chesapeake Bay on the USRC *Miami*, in May of 1862, which carried President Abraham Lincoln and other Federal officials into a combat zone to assess federal military and naval forces and then order them into combat. The *Miami* supported the U.S. Navy and U.S. Army in putting Union troops ashore in Virginia[4] and the attack on Norfolk.

Ostrom and Galluzzo surveyed the career of Captain Charles F. Shoemaker (USRCS), who matriculated from the U.S. Naval Academy as a midshipman and transferred to the Revenue Cutter Service, where he earned a commission in 1860. Third Lieutenant Shoemaker was assigned to the USRC *Lewis Cass*, which was home-ported in Mobile, Alabama, under the command of the pro-secessionist commander James J. Morrison (USRCS), who transferred the cutter to Confederate officials in 1861. Lt. Shoemaker and his Federal crew members were forced to trek across the Confederate States to reach the safety of Union territory. During the Civil War Shoemaker served on cutter escort missions off the East Coast, in the port of New York, in what would later be termed "port security" but in his time was called "guard duty." Shoemaker resigned his commission after the Civil War to pursue a business career and then returned to the USRCS in 1868. He would command different revenue cutters (USRC *Seward*, USRC *Washington*, USRC *Hudson*), and serve as a district Life-Saving Station inspector for the U.S. Life-Saving Service. United States Treasury secretary John G. Carlisle appointed Captain Shoemaker to command the U.S. Revenue Cutter Service from 1895 to 1905.[5]

James M. McPherson, the eminent historian of the Civil War, divided his naval analysis into two themes: "Blockade and Beachhead: The Salt-Water War (1861–1862)" and "The River War in 1862."[6] The Union Blockade of the South presented the U.S. Navy and U.S. Revenue Cutter Service with formidable logistical and tactical challenges. McPherson described the CSA as controlling 3,500 miles of coastline, ten major ports, 200 navigable inlets, and numerous river mouths and bays that smaller vessels could traverse and privateers and raiders could sail in and out of. The USN had just over 30 ships on patrol in 1861 and only two major

Rear Admiral Charles F. Shoemaker served as captain-commandant from 1895 to 1905. Prior to that, Third Lieutenant Shoemaker served in the Civil War. As chief of the Revenue Marine Division, Shoemaker completed the transformation of the revenue cutter fleet from sail to steam propulsion.

bases south of the Mason-Dixon Line: Hampton Roads, at the mouth of the James River, and Key West. Confederate military forces captured the U.S. Navy base at Norfolk, Virginia, early in the war. The effectiveness of the Union Blockade was problematic and debatable, but it had significant politico-economic, social, and psychological effects. Most Rebel blockade-runners penetrated and evaded the blockade at the beginning of the war. About half the attempts to bypass it were successful in 1865. Southern civilians, politicians and military and naval leaders reported significant hardships and deprivations in their efforts to obtain essential supplies from Europe and the West Indies.[7] Despite the gaps and shortcomings in the Union naval blockade, and the debatable conclusion that it "won the war," McPherson concluded, "It did play an important role in the Union victory. Although Union naval personnel constituted only 5 percent of the Union armed forces, their contribution to the outcome of the war was much larger."[8]

In 1861 and 1862 the Union river wars were strategically insignificant, but from 1862 until the end of the war the rivers and interior coastal waters and bays were important theaters of operation. On the upper Mississippi River, Cairo, Illinois, became a military and naval base. Joint U.S. Army-Navy operations were essential and facilitated good communications, transportation and cooperation between the Federal armed forces. Initially, the U.S. Army controlled the amphibious operations, and the War Department built the first river gunboats. Heterogeneous crews manned the vessels and included volunteer civilian riverboat men, steamboat machine engineers, and experienced riverboat pilots. By late 1862 the U.S. Congress had put the river naval squadrons under the control of the U.S. Navy to manage combat operations and transport troops and supplies.

Maritime warfare was waged by U.S. Navy and U.S. Revenue Marine ships, boats and other watercraft against fewer but dangerous Confederate naval and military forces and the formidable Rebel land-based forts. Among the strategic Southern ports, cities, and infrastructure to fall to Federal naval and military forces were New Orleans, Vicksburg, and the ships, boats, bastions and mines at Mobile Bay.[9] McPherson ably traced the logistical and tactical cooperation between General Ulysses S. Grant and his U.S. Army colleagues, including the reluctant, arrogant and stubborn General George McClellan, and leading U.S. Navy officers like Admiral David G. Farragut and Commander David Dixon Porter; the leadership of the United States War and Navy Departments; and President Abraham Lincoln. Even Secretary of War Edwin M. Stanton, thought of as arrogant and nasty by many of his critics, cooperated with military and naval leaders to craft strategies and achieve wartime objectives. Politics, personalities, and egos had to be confronted. Professor McPherson chronicled some of them. Considering the potential for professional clashes within and between the armed services and the complex planning involved in the operations, McPherson concluded that the interservice cooperation exceeded inevitable rivalries and resulted in the Union victory over the Confederate States of America resulting in the reunification of the United States.

In his classic book, *Ordeal by Fire*, James M. McPherson offered other insights in his analysis of Civil War naval strategies, tactics, and missions.[10] He used historical geography to explain offensive and defensive military and naval tactics, strategies, and the relative advantages and disadvantages physical, cultural and economic geography posed to the opposing forces. Between the Federal capital at Washington, D.C., and the Confederate capital at Richmond, Virginia, lay at least six major rivers and the numerous tributaries that formed the transportation arteries, barriers, and advantages to the belligerents. The Mississippi River, wrote the author, "was an arrow thrust into the heart of the South," and the Tennessee and Cumberland rivers served as Federal "highways of invasion"[11] into the Confederate States of Alabama, Tennessee and Mississippi. Union naval forces possessed certain transportation and other infrastructure advantages. Confederate railroads, bridges, roads, horse-drawn supply wagons and artillery utilized and facilitated Union logistics, tactics and supply operations but were vulnerable to seasonal weather conditions, enemy guerrilla units, civilian and military espionage, and mounted cavalry troops.[12]

Admiral David Farragut (USN) conducted extensive and successful naval battles against Confederate land and maritime forces on the lower Mississippi River and Mobile Bay, Alabama. In those conflicts, Farragut preferred auxiliary (steam and sail/wind) powered combat ships for maneuverability and fuel economy. He preferred his wooden-hulled ships against Confederate ironclads, McPherson asserted, because enemy shot might penetrate and travel completely through a wooden hull without sinking the vessel. But enemy shot and shore-side artillery could damage and sink iron-hulled vessels. Although Adm. Farragut's flagship was the vaunted USS *Hartford*, McPherson described how the admiral favored another warship in his line of combat vessels, a fast-cruising, three-masted, 23-gun auxiliary sailing sloop, the USS *Pensacola*. In heavy combat during the run between the forts of the lower Mississippi on the way to New Orleans, the *Pensacola* survived numerous "direct hits."[13]

McPherson described Southerners as a "martial" people, not a "maritime" population. The North had the edge in trained naval officers and merchant seafarers who did not need to learn their professional specialties from an elementary level, as Northern military volunteers did. The U.S. Navy and U.S. Revenue Cutter Service were led by professional, commissioned officers, not, as was true in the volunteer U.S. Army and state militia ranks, political appointees and officers voted into their commands by loyal, regionally cohesive troops. The North also had the advantage in industrial production. Ships that were not built in Northern shipyards were purchased from civilian and merchant marine sources. The South had to build and convert combat ships and ironclads and order vessels from overseas while European nations were still sufficiently neutral in the American Civil War, before diplomatic and economic pressures ended those sources for the Confederacy.

Confederate navy secretary Stephen Mallory was innovative enough to initiate the use of "torpedoes," now called "mines" in naval parlance, and formidable ironclad vessels that took a toll on Union warships and merchantmen. The South manufactured and manned the submarine *H.L. Hunley*, which sank a Union warship in Charleston Harbor in 1864 but fell victim to the attack and sank with all hands. Successful civilian privateers, blockade-runners, and raiders aided the Confederate cause. The Confederate ironclad CSS *Atlanta* was typical of the South's armored warships. The vessel featured an iron prow situated below the water's surface and designed to tear into the wooden-hulled Yankee warships below the waterline, causing them to sink. Federal sailors finally captured the *Atlanta* on the Savannah River in Georgia. The South built the first American iron warship, having converted the sunken USS *Merrimack* into the formidable and successful CSS *Virginia*, which battled the iron-hulled USS *Monitor* to a draw. The Union naval fleets and squadrons had the wartime advantage of skilled, well-trained crew members and officers, the administrative talent of U.S. Navy secretary Gideon Welles, and, as McPherson described him, the "dynamic" and knowledgeable assistant navy secretary Gustavus V. Fox.[14]

Professor Dean Jobb of the University of King's College (Halifax, Nova Scotia) described a unique episode in the history of Confederate raiders in the American Civil War. Professor Jobb brings geographic and historical perspectives to what he called "The Confederacy's Last Great Raid on Union Shipping."[15] He chronicled the late-night passage of the "sleek steamer" CSS *Tallahassee*, which sailed out of the Wilmington, North Carolina, harbor in the late evening darkness of 6 August 1864. Two U.S. gunboats converged across the treacherous sandbars and shoal waters to confront the Rebel raider, but the experienced *Tallahassee* helmsman passed between the vessels at such close quarters that the skipper, Captain John Taylor Wood (CSN), said the respective crews could have thrown food onto the decks of the vessels. Union gunners fired at the Confederate raider but scored no direct hits. The Rebel steamer eluded those and two other Federal warships and escaped into deep-blue ocean waters. Captain Wood ordered his deck guns to remain silent so Yankee sailors might conclude his was a civilian privateer, an ordinary blockade-running merchant vessel and not a Confederate States Navy gunship. The CSS *Tallahassee* commenced a twenty-day mission of commercial ship destruction that Jobb described as the South's "last great raid on Northern shipping that would test the already strained relations between London and Washington, catch the Union Navy off guard, and spread panic among merchant ship owners."[16]

Captain Wood (CSN) had a Minnesota connection, having been born in 1830 in the frontier state at the strategically located Fort Snelling in what is now St. Paul. John Taylor Wood's father, Robert Wood, was stationed in the U.S. Army at Ft. Snelling as an assistant surgeon. General Zachery Taylor, the future president of the United States, was the commander of the fort. His daughter Anne Taylor was Captain Wood's mother. To continue the kinship complexity, and future political alliance, a young U.S. Army officer named Jefferson Davis, who would become the president of the Confederate States of America, married one of General Taylor's other daughters, Sarah. The sisterhood connection of the Taylor family made the future commander of the CSS *Tallahassee* the nephew of Jefferson Davis. That explains one of the reasons why Captain Taylor Wood, with mixed emotions and some reluctance, joined the Confederate Navy.

In 1847, at the age of 16, Wood earned an appointment as a midshipman in the U.S. Navy, and was stationed as a gunnery officer during the Mexican-American War (1846–1848). After the war, John Taylor Wood lived at the White House for a time with President Taylor. Wood expanded his naval career when he was appointed to serve at the U.S. Naval Academy at Annapolis, Maryland, as a gunnery instructor. Despite his ties to the U.S. Navy and the Union, John Taylor Wood and his brother joined the Confederacy and fought for his uncle, Jefferson Davis. His parents and other relatives stayed with the Union.[17] As mentioned, Captain John Taylor Wood of the CSS *Tallahassee*, in the words of the Canadian Professor Dean Jobb, "was the grandson of former United States president Zachery Taylor and a nephew of Confederate president Jefferson Davis."[18] The family-splitting nature of the Civil War and the mixed experiences and loyalties of its participants is graphically illustrated in the family history of Captain John Taylor Wood of the United States Navy and Confederate States Navy.

Captain Wood was the gunnery officer on the CSS *Virginia* in the successful March 1862 attack at Hampton Roads on the Federal ships USS *Congress* and USS *Monitor*. Captain Wood scored direct hits, participated in the destruction of several other Union commercial and naval vessels, and met with and advised President Jefferson Davis on naval tactics and strategy. Then, toward the end of the war, Captain Wood planned to use the fast (15 knot), British-built, twin-engine, iron-hulled, 220-foot CSS *Atlanta*, renamed CSS *Tallahassee*. The vessel had a 120-crew complement, three guns, and a 100-pound (weigh of shot) deck cannon and successfully breeched the Union blockade off Wilmington, North Carolina. Captain Wood and his crew also seized Union commerce vessels, took essential equipment, provisions and other supplies, burned vessels to the waterline, and ordered surviving ships to take passengers and crews on board and escort them to the nearest ports. On just one day, for example, the *Tallahassee* captured six vessels off New York City and Long Island. In addition, Wood captured fishing boats and trade vessels.

By late August 1864 the CSS *Tallahassee* had already captured several dozen ships. But, with a diminished coal supply and Union warships closing in, Captain Wood decided to sail his ship to Halifax, Nova Scotia, the nearest neutral port. A British Royal Navy Squadron was stationed at Nova Scotia. Pro-Southern sympathizers, blockade-runners, and nefarious operatives found security and hospitality in the generally anti-American, pro–British port city. Yet, British and Canadian officials limited the Confederate warship's supply and rest mission so as not to antagonize the United States or have Canada's neutral status and commercial opportunities challenged,[19] even as pro–Rebel merchants and a Halifax physician (Dr. William J. Almon) cheerfully met the support, supply, recreational, and repair needs of the CSS *Tallahassee* and its crew. On 20 August 1864 Captain Wood met his time limit in Halifax and planned to sail back to Confederate waters before pursuing U.S. Navy warships trapped him. Sailing in dangerous shoal waters and darkness, the *Tallahassee* commander and crew reached Wilmington on 25 August 1864, after a brief clash with Union warships. Professor Jobb cited the Canadian historian Greg Marquis, who concluded that Captain Wood boosted Confederate pride and morale, demoralized and frightened American merchants and shippers, and, "embarrassed the Union Navy."[20] After the Union capture of Wilmington and the surrender of the CSA to U.S. authorities, Captain Wood and other Rebel sailors and merchants returned to Nova Scotia to establish commercial enterprises, and Wood formed a successful shipping firm. He died in Halifax at the age of 73, in 1904.[21]

William M. Fowler, Jr., wrote a compelling history of America's Confederate and Federal navies in the Civil War. His book describes the two navies as operating "under two flags."[22] Professor Fowler has published numerous articles and books on the Civil War and has taught history at Northeastern University in Boston, Massachusetts. His scholarly studies of the Civil War navies led him to conclude that, while U.S. Navy secretary Gideon Welles had to "organize and rally his fleet" after Fort Sumter, Confederate States Navy secretary Stephen Mallory "had to create a navy where none existed."[23] While the Union had a naval and maritime heritage, the South lacked such a tradition. Fowler contended that President Jefferson Davis, a graduate of the U.S. Military Academy at West Point, New York, thought land warfare would secure victory for the South and that the Confederate Navy was simply an army auxiliary force. Mallory's insistence on a significant naval defensive and offensive buildup and role did modify the priorities of Davis. Mallory began to construct warships at Memphis, New Orleans, and Norfolk as well as to purchase commercial vessels and fishing boats for naval reconstruction. Fowler described the competitive strategies of the Confederate and Union navies. The Union hoped to confine Confederate raiders and warships to Southern ports with a strong blockade. The Confederate strategy was to break the Union naval blockade with warships and commerce raiders and force the U.S. Navy and Revenue Marine to come out to sea to engage Rebel vessels in order to protect Northern commerce.[24]

Professor Fowler chronicled the industrial, economic, demographic, infrastructure, and naval and military dominance of the Federal government over the Confederacy, and the eventual defeat of the latter. After General Robert E. Lee surrendered to Union forces on 9 April 1865, President Jefferson Davis fled Richmond but was captured by Federal troops and imprisoned. Confederate Navy secretary Mallory fled Richmond but was captured and imprisoned until freed by Federal officials in March of 1866. He died from a heart attack in 1873. After President Lincoln's assassination, U.S. Navy secretary Welles would serve President Andrew Johnson through the failed attempt by Radical Republicans to drive Johnson from office. After the Johnson administration, Welles retired to write his memoirs and articles and publish his diaries, the latter a task never completed. Gideon Welles passed away in 1878.[25]

As exciting and destructive as the war of Southern raiders against Union commercial vessels around the globe was, Fowler concluded that the victories were less impressive than the image. While successfully displacing and destroying hundreds of Union merchant and whaling vessels, and the increase in maritime insurance rates, the ability of the United States to engage in commerce and also wage war was not significantly diminished. Navy Secretary Welles was forced to dispatch some of his fleet to search for raiders, but that mission did not divert Union warships from administering the successful and punishing blockade of Southern ports.Nonetheless, Professor Fowler concluded, the Rebel raiders and Confederate ironclads contributed adventure and innovation to maritime history and to their "Lost Cause."[26] However, Fowler continued, "The Union Navy was a powerful partner with the Northern Army, and its share in the Battle for the Republic should not be forgotten."[27]

James McPherson devoted one of his several Civil War books to the role and significance of the Confederate and Union navies in his book *War on the Waters*,[28] in which he covered the significance of the Union blockade, Confederate nautical ingenuity, American blacks in the Union navy (blacks also served in the U.S. Revenue Marine), and the shortcomings and strengths of Confederate and Federal naval commanders in the battles conducted on the oceans, bays, inlets, harbors and rivers of North America. The author cited the opinion of a Confederate navy midshipman and historian regarding the significance of the Union blockade of the South, which "shut the Confederacy off from the world, deprived it of supplies, and weakened its military and naval strength."[29] McPherson stated his own nuanced conclusion about the Union naval blockade of the South: "To say that the Union navy won the Civil War would state the case much too strongly. But it is accurate to say that the war could not have been won without the contribution of the navy."[30]

In terms of the contributions and significance of the U.S. Revenue Cutter Service to naval combat, McPherson offered insights into the missions, exploits, achievements, and unfortunate fate of the famed RC *Harriet Lane*, which supported Lincoln's logistical plan to use naval and merchant vessels to resupply U.S. troops stationed at Fort Sumter in Charleston Harbor.[31] That effort, of course, on 12 April 1861 initiated the Confederate shelling of the fort and the commencement of the Civil War. McPherson takes the reader to Galveston, Texas, when three companies of U.S. troops arrived to support the five naval gunboats at the port. In the early morning hours of New Year's Day 1863, four Confederate steamboats carrying protective cotton bales and one thousand sailors and soldiers launched an attack. Two of what McPherson termed "the cotton-clads" rammed and hemmed in the vulnerable side-wheel steamer USS *Harriet Lane* and engaged the 100 Federal crew members in fierce deck combat, during which Commander Jonathan Wainwright (USN) was killed. McPherson quoted a Confederate surgeon who witnessed the battle: "Our boys poured in, and the pride of the Yankee Navy was the prize of our Cowboys."[32]

An understanding of historical geography is essential for students of the naval and military campaigns of the Civil War. Former U.S. Naval Academy historian Craig L. Symonds brings that understanding to his several Civil War histories, most particularly in his classic, *The Civil War at Sea*.[33] Symonds offers a compelling narrative, enhanced by the mapmaking skill of cartographer William J. Clipson. Numerous maps illustrate the terrestrial and maritime domains of battles between Confederate and Union military and naval forces on the Mississippi River, Mobile Bay, Charleston Harbor, the Atlantic and Mexican Gulf maritime regions, and Chesapeake Bay. Photographs and illustrations of Confederate and Union ships and Civil War political, military, and naval leaders add to the instructive mix. Professor Symonds explains leading events, strategies, tactics, and outcomes in the context of graphic illustrations, personalities and incidents. Symonds assesses the ships, guns, and technologies of the naval war; the blockade and blockade-runners; attacks on commercial and naval vessels and forts and harbors; and the courage of the naval personnel who had to confront the horror and carnage of war. He also describes the technology and tactics of wooden- and iron-hulled vessels, auxiliary (sail-wind and steam) power, and the skills officers and crew members had to master in a new and rapidly evolving maritime world. The author assesses the backgrounds and competencies of Confederate and Federal political and naval leaders as well, describing Stephen Russell Mallory as a ""hard-working and competent Confederate Secretary of the Navy who advocated iron-armored warships to compensate for the Union's numerical superiority," one who initiated "the establishment of a Confederate Naval Academy."[34] About the U.S. Navy secretary he had this to say: "Lincoln appreciated Gideon Welles for his candor and loyalty,"[35] and of Captain Raphael Semmes (CSN), the commander of the Confederate raider, CSS *Alabama*, he stated, "He ravaged Union shipping on three oceans and provoked terror in the hearts of Union shippers"[36]; Commander James Dunwoody Bulloch (CSN): "The mastermind behind the ship acquisition program in England"[37]; Captain Samuel Francis Du Pont (USN): "The Union's first naval hero presided over the Strategy Board ... commanded blockading squadrons ... and won the North's first victory at Port Royal, South Carolina ... but earned the disapproval and animosity of his civilian masters."[38] And Symonds's description of the Confederate raider CSS *Shenandoah* (Lt. Cmdr. James Iredell Waddell, CSN) sailing was that it, "under bare poles [masts] among the ice floes of the North Pacific, fired the last shots of the Civil War."[39]

The Confederacy lost the war because it could not match the Union's industrial productivity, wealth, demographics, military and naval equipment, resources, and supplies, according to Symonds, who concluded with a statement on the significance of Union naval dominance: "Though scores of small blockade runners passed through the blockade, it contributed to a growing sense of isolation in the South. Naval forces did not determine the outcome of the Civil War. The North would have won the war without naval supremacy. But naval forces affected the war's trajectory and very likely its length, and that, in the end, was important enough."[40]

The conclusions of Professor Symonds are measured and compelling. But it might also be argued that the influence of the Union naval war was essential to victory. The powerful and extensive U.S. Navy and U.S. Revenue Cutter Service personnel, ships, and combat split the South into isolated regions, forcing the Confederacy to use land routes and to have to substitute for the resources and supplies lost to Union military and naval forces on land and sea. The supplies, arms and equipment eventually failed to arrive from Europe, and the export revenues and trade the Confederacy lost dried up overseas and added immeasurably to the South's economic problems and ability to achieve its objectives.

13

Policing the Alaska Frontier (1867–1915)

In September 1860, seven months before the Civil War began and before the president-elect, Abraham Lincoln, appointed him secretary of state, William H. Seward proclaimed his foreign policy vision in a speech in "the central northern highlands," as historian Walter Star described the geography of St. Paul, Minnesota. On a day of brisk fall weather, Seward said he could visualize the future, looking north into what was then Russian America (Alaska), where future settlers would "build up outposts all along the coast, even up to the Arctic Ocean; outposts of my own country."[1]

The quest for Alaska and its eventual purchase would significantly expand the maritime domain and mission responsibilities of the U.S. Revenue Cutter Service and its successor agency, the United States Coast Guard. The expansion of U.S. jurisdiction into the Arctic and polar regions would have enormous significance in subsequent national defense, search and rescue, environmental protection, sustainable resource exploitation, and other issues from 1867 to the present day. The duties, missions and responsibilities of the variously named U.S. Revenue Marine, Revenue Cutter Service, and U.S. Coast Guard have expanded as the Service has responded to changing times along with limited budgets.

Captain Leonard G. Sheppard served as chief of the U.S. Revenue Marine Division and Commandant of the U.S. Revenue Cutter Service from 1889 to 1895, during the early stages of the Bering Sea patrols in Alaskan and Arctic waters. Sheppard facilitated the training of engineering officers to manage increasingly sophisticated cutter steam power and mechanical technology. Commandant Sheppard also established the position of commander of the Bering Sea Fleet on the Bering Sea Patrol. Captain-Commandant Sheppard defined the significance of the expanded cutter missions when he referred to the Service as "the maritime constabulary of the nation."[2] The background of the expansion of the cutter missions with the acquisition of Alaska as a territory can be traced back to the diplomatic machinations and far-sightedness of William Henry Seward, the New York lawyer, state legislator, governor, senator, diplomat, and domestic and foreign policy advisor to presidents Abraham Lincoln and Andrew Johnson.

As Lincoln's secretary of state, Seward advised the president on domestic political issues and policies, as well as foreign policy. Seward continued in the State Department in the administration of Andrew Johnson and completed the negotiations for the purchase of Alaska from Russia in 1867. The negotiations were between Seward and Baron Eduard de Stocky, the Russian envoy from the tsarist government to the United States. Ambassador Stocky had expressed his concern and counsel to the Russian foreign minister

Captain Leonard G. Shepard served as Chief of the Revenue Marine Division from 1889 to 1895 and stationed the training ship USRC *Chase*, a cutter he had commanded, at Arundel Cove in Baltimore to serve as the School of Instruction vessel.

that perhaps Russia should sell its Arctic and Bering Sea maritime and terrestrial domains to the United States before the Americans moved in and took the region by demographic dominance, and perhaps war, as the United States had done in the former Mexican Southwest and West.[3]

Reluctant and dubious members of the United States Senate, after much debate, ratified the Treaty of Annexation and Purchase of Alaska, a territory larger than the state of Texas, for $7.2 million on April 9, 1867. President Andrew Johnson signed the treaty on May 28. Critics initially denounced "Seward's Folly," until the Alaskan gold rush of 1896 and subsequent discoveries of abundant maritime and terrestrial resources vindicated Secretary Seward.[4] The territory of Alaska would prove to be geopolitically and strategically significant. Soon after the ratification of the Treaty of Annexation, President Johnson recommended that Congress provide for the military occupation and governance of the territory of Alaska as part of the United States. It would be administered and policed by the U.S. Army, the U.S. Revenue Cutter Service, the U.S. Treasury Department and the Collector of Customs. The steam- and sail-powered USRC *Bear* patrolled Alaskan waters for almost half a century, going north in the spring from the Pacific Northwest to take government personnel, medical services, and mail, and provide "law and order" to the region. It also took Siberian reindeer to starving Inuit Eskimos[5] in extraordinarily complex logistical missions.

The 198-foot USRC *Bear* operated in the Arctic and Bering Sea in a plethora of multi-mission assignments and had a reinforced hull that sustained its voyages in the dangerous, icy seas of the polar region.[6] Coast Guard historians Truman R. Strawbridge and Dennis L. Noble described the strengths and weaknesses of Michael A. Healy, the commander of the *Bear*, his courage, seamanship skills, extraordinary mission challenges, heroism, alleged drunkenness, cruelty in disciplining his enlisted personnel, court-martial, suspension from and return to active duty, and concern for and transfer of reindeer herds to the indigenous Inuit people.[7] The U.S. Coast Guard posthumously honored the famed cutter captain by launching the polar icebreaker and scientific exploration platform USCGC *Healy*, in 1998.[8] The *Healy* missions became increasingly significant in oceanographic research, search and rescue, and the establishment of a United States naval presence in the increasingly geostrategic and economic significance of Arctic waters. In their extensive study of Captain Michael A. Healy's life, seamanship skills, courage, and contributions to Arctic exploration and assistance to the indigenous Inuit people, Strobridge and Noble chronicled the contributions of the U.S. Revenue Cutter Service.

The authors balanced their compliments and criticisms of the controversial captain as they traced his extraordinary life, bravery, and toughness through his senior years, including hazardous SAR missions to find and save iced-in whaling ships; his transfer of reindeer from Siberia to Alaska in a social engineering project to provide food to starving Inuit; the faith Native Americans had in his intentions and humanitarianism; the irony of his attempt to prevent the devastating impact of alcohol on the Inuit while he himself over-indulged in intoxicants; the controversial revelations regarding his African-American heritage; and the political influence of his brother James, a Roman Catholic priest.[9] Strobridge and Noble summarized the career of the controversial cutter man: "There may be questions surrounding the revenue captain, but one fact remains: Capt. Michael A. Healy was one of this country's great Arctic navigators, an enigmatic man who still has not received the place in history he deserves, but [with the 1998 launching of the polar icebreaker *Healy*], it seems fitting that the name of Michael A. Healy will again be seen in the waters of the far north."[10]

Historian Paul H. Johnson published an extensive account of one of the most dangerous, heroic, skillful and successful rescue missions etched in the legendary history of the U.S. Revenue Cutter Service: The Overland Mission of 1897–1898 across the Arctic wilderness of northern Alaska, an expedition commenced by the USRC *Bear.* Johnson published his research in the September/October 1972 issue of the U.S. Coast Guard Academy Alumni Association periodical, *The Bulletin* (vol. 34, no. 5). A fleet of whaling vessels in Arctic waters had been iced in and faced isolation and starvation, given the unlikely possibility of being reached by any potential rescuers navigating the frozen Bering Straits to the north coast of Alaska, where the whalers were marooned. "News of the whalers' peril was brought out of the Arctic by the whale ship *Karuk.*"[11] The San Francisco Chamber of Commerce sought the assistance of President William McKinley (presidency 1897–1901). McKinley, later assassinated, would be noted for the Overland Expedition by the USRCS and the involvement of the United States in the Spanish American War of 1898.)

President McKinley responded quickly, given his familiarity with a scientific expedition that had been lost in the previous decade, when the Arctic scientific expedition of Lt. Adolphus Greely (U.S. Army), with just seven sur-

The Revenue Cutter *Bear* was a leading element of the Bering Sea Patrol after the purchase of Alaska from Russia by the United States in 1867. The *Bear* crews carried out law enforcement, criminal justice administration, the transportation of supplies and federal officials, and the enforcement of federal fish and game laws, search and rescue, and icebreaking. In 1897 and 1898 members of the *Bear* crew carried out the dangerous Overland Expedition rescue of iced-in whaling crews off Pt. Barrow, Alaska.

U.S. Revenue Cutter *Bear* leading SS *Corwin* (a former revenue cutter) through polar ice toward Nome, Alaska, in 1914.

vivors, was belatedly rescued in 1884 by the USRC *Bear* under the command of Captain Winfield Scott Schley (USRCS) and two U.S. Navy vessels. McKinley ordered Treasury Secretary Lyman J. Gage to assign the USRCS, and Captain Francis Tuttle, then commander of the RC *Bear*, to study the circumstances, suggest methods of operation, and commence the rescue of the 265 crewmembers aboard the iced-in whalers. Because it would be impossible to navigate the *Bear* through the winter ice, it was deemed necessary to plan an overland expedition of approximately 1,500 miles, using experienced cuttermen, indigenous native guides, dogs, sleds loaded with food, and experienced Alaskan herders to guide deer herds on the hoof for food to the site.

Captain Mike Healy, the strict and controversial commander of the famed Revenue Cutter *Bear*, on the Bering Sea Patrol. Healy's acquisition and transportation of reindeer to the starving Inuit people was an extraordinary logistical achievement.

The chief of the Revenue Marine, Commandant Charles F. Shoemaker (USRCS), who served in that capacity from 1895 to 1905, chose the RC *Bear* and its seasoned Arctic crew for the volunteer rescue mission. The 200-foot, reinforced wooden-hull *Bear*, powered by steam and sail (wind) with one screw, was a forerunner of Coast Guard icebreakers and ideally suited for that mission in the extreme November weather and maritime conditions. The *Bear* crew had just docked in Seattle but quickly prepared the vessel with stores and other equipment for the polar mission. Ice conditions prevented the cutter from advancing beyond the southern part of the Bering Sea. The volunteer rescuers then proceeded on land to St. Michael, a village where more supplies would be acquired for the trip to the next site, Cape Prince of Wales. Then the rescue team would be tasked with acquiring a reindeer herd and proceeding to Point Barrow at the farthest point of the north coast of Alaska.

By that stage, mid–December 1897, the brave volunteer rescue crew consisted of among the most experienced and courageous men of the Revenue Cutter Service: Second Lieutenant E.P. Bertholf (who would subsequently be appointed the captain-commandant and commodore of the U.S. Revenue Cutter Service, and then the U.S. Coast Guard (1911–1919); surgeon and physician Dr. S.J. Call; and First Lieutenant David H. Jarvis, who was fluent in the Eskimo/Inuit languages. All of their talents would be essential for survival on the trek and in the management and care of the surviving whaler crew members upon contact.[12] Ice and heavy seas almost swamped the boats the lifesaving crew used to get themselves and the supplies ashore. The *Bear* would meet the cuttermen again eight months later, when the ice pack would be diminished at Point Barrow. The crew trekked 1,500 miles in temperatures of minus 30 to minus 60 degrees Fahrenheit in the dark and gloom of the long Arctic winter wilderness, with four hundred reindeer acquired from Eskimos and a missionary who accompanied the crew. Some Eskimos had warned against the trip, which eventually took 100 days over rough, dangerous ice and snow mounds, crevasses, mountain passes, frozen tundra, and blizzards. Isolated Inuit groups, Eskimo guide and herder Charlie Artisarlook, and missionary and reindeer herder W.T. Lopp kept the men on track and alive. The instincts of the 200 reindeer and dogs (which suffered terribly as well) were also instrumental in the successful trek.

In March 1898 the rescuers spotted the rigging and spars of the western-most whaling vessel, which was banked up on a snow ridge. Some sailors were living on the ships, others in wretched huts onshore. The starving men were suffering from scurvy in an environment of sickness, filth and death. The rescuers immediately got busy. Over time, consumption of the fresh meat alleviated the scurvy. Newer wood shelters from the wood of the disabled ships provided more adequate and warmer shelters. The men also played baseball games on the ice to provide some exercise and counter the depression of both the whalers and the rescuers.[13]

The USRC *Bear*, under the command of Captain Tuttle, eventually traversed the melting fields of ice and reached Point Barrow on 28 July 1898. The cuttermen would later earn the gratitude of the president and the nation and Congress awarded the rescuers gold medals. The Overland Expedition was a success. Most of the whaling crew members survived and traveled back to Seattle on the *Bear*. The meticulous, articulate diary kept by Lt. Jarvis provided a vivid history of the logistics and events of the expedition. As the historian and author Paul H. Johnson perceptively and eloquently concluded in his research, the Overland Expedition and "the stirring Jarvis journal stands as a testament to human courage; a reminder of a time when a small band of dedicated mariners brought succor and justice to the Alaskan frontier. Today, Coast Guard vessels and men and women cruise Alaska's dangerous waters, always ready to aid those in distress."[14] As the editor of Johnson's publication noted, "The Service's [then] newest 378-foot high endurance cutter, the CGC *Jarvis*, was commissioned at Honolulu, Hawaii on 4 August 1972, [and] named in honor of the Overland Expedition leader, Captain David H. Jarvis." The ceremony was attended by the captain's 77-year-old daughter, Anna Jarvis.[15] The 4 August date was especially appropriate, as that is the month and day of the birthday of the U.S. Revenue Marine/Revenue Cutter Service and United States Coast Guard in 1790.

Irving H. King, professor of history at the U.S. Coast Guard Academy in New London, Connecticut, described the living conditions of the U.S. Revenue Cutter Service rescue personnel on the Overland Expedition. They preferred living in their own tents in sleeping bags instead of the Inuit dwellings. The sleds that came on the RC *Bear* from Seattle were big, tough, bulky, and unwieldy. Dog teams were fed chopped-up frozen seal meat and slept outside in storms. The USRCS rescue team drank tea and ate beans, hard bread, bacon, crackers, and fried flapjack cakes made of water and flour and fried in the heating devices the team carried. During breaks, the rescuers smoked tobacco. Alexis Kalenin, a Russian trader, guided the team for part of the mission and provided sleds and dogs for the daily 20- to 50-mile treks, the distance depending on weather conditions. Inuit villagers along the way were "kind, hospitable, and generous," but their "huts were crowded, filthy, and poorly ventilated," and the natives were "none too healthy."[16]

Sometimes good speed was made along the snow-free sections of the frozen Yukon River. When the team came upon towns and villages, and in one case an iced-in steamer, they purchased dogs, supplies, and clothing. Incredibly, Lt. Jarvis met an Inuit woman he knew from the far north village of Point Hope. She and her husband were escorting George F. Tilton, a rugged survivor of the whaling expedition who had traveled hundreds of miles. His Inuit comrades were taking Tilton south to St. Michael to get assistance. Tilton informed Lt. Jarvis of the horrible conditions facing the whalers and what the revenue cutter officer already knew, the names of the marooned whaling vessels: the steamers SS *Belvedere, Freeman, Rosario, Newport, Fearless,* and *Wanderer,* and the four-masted schooner *Jeannie.* The whaling ships were sail- and steam-powered.[17]

The cuttermen (officers and enlisted personnel) who sailed in Alaska-Bering Sea-Arctic-Polar waters and performed rescue missions on land and sea deserve the accolades and honors for the dedication, skill, and danger involved in their work. Strobridge and Noble described the duties and honored the Revenue Service men but lamented the fact "the officers and crews had to suffer long hours in an extremely hostile environment, standing lookout watches in the ice pack ... when the polar winds howl across the frozen sea and cut into your very soul. Officers might receive recognition for their work, but those [enlisted crewmen] who sailed before the mast earned very little, if any, acknowledgment."[18]

In his classic history, *The United States Coast Guard, 1790–1915: A Definitive History* (1949), Captain Stephen H. Evans (USCG), writing only about 50 years after the 1897–1898 Overland Expedition, offered valuable insights. He explained how the people of America followed the tragedy and the rescue attempt over the wires and in the press. The primary source for the detailed operations and history of the mission was the official *Report of the Cruise of the U.S. Revenue Cutter* Bear *and the Overland Expedition for the Relief of the Whalers in the Arctic Ocean,* which included the masterful log, chronicled by First Lt. David H. Jarvis (USRCS), the leader of the mission.[19] Evans studied the official report and shared the riveting descriptions of the expedition events in the prose of Lt. Jarvis, including this passage: "Our mountain climbing for the day was not over, for there was still another portion of the range to be crossed even higher and steeper than the one we had just come over. In the course of time, after much tugging and pushing of sleds and urging of dogs, we reached the summit, where we found ourselves in the midst of a furious storm of wind and snow so thick, the natives could not decide upon the proper direction. There was some danger of our taking the wrong course and going over a precipice into the sea."[20] Evans described how Jarvis and his team finally reached the stranded and starving whaling crews, and how the revenue officer managed and administered health treatments and food supplies and adjudicated disputes between the ill and unhealthy whalers and the Inuit. Dr. Call visited and treated the crews still aboard the iced-in whaling ships.[21] The RC *Bear* finally was able to access Point Barrow in August 1898. Stephen Evans summarized the finale: "The cutter helped the whaling ships out of the ice, took the Overland Expedition and stranded whalers on board, and headed home."[22]

In his welcome to the members of the expedition, the crew of the RC *Bear,* and the rescued whalers, President William McKinley addressed Congress as follows: "I commend this heroic deed to the grateful consideration of Congress and the American people. The year [1898] just closed has been fruitful with noble achievements in the field of [the Spanish-American War], and while I have commended to our consideration the names of heroes who have shed [glory] upon the American name in valorous contests and battles by land and sea, it is no less my pleasure to invite your attention to a victory of peace."[23] Congress, in the act of June 28, 1902, awarded special gold medals to revenue cutter service officers Second Lt. E.P. Bertholf, First Lt. D.H. Jarvis, and Dr. S.J. Call for "heroic service rendered."[24]

Between the end of the Civil War in 1865 and the commencement and end of the war between the United States and Spain (1898), the U.S. Revenue Cutter Service stayed busy with maritime missions and duties in Alaskan waters, the Aleutian Islands, and the polar Arctic. The establishment of U.S. sovereignty in the sparsely settled Alaskan maritime regions was a challenging mission in the vast region of approximately 40,000 white settlers and indigenous Americans of Inuit and Indian heritage. Some Native Americans were hostile and aggressive. Most were not. From the time of the U.S. purchase of Alaska and over the following decade U.S. authority was symbolized by scattered U.S. Army troops in isolated compounds, a collector of customs in Sitka on the Alaskan panhandle immediately west of British Columbia in Canada, and the cruising revenue cutters, said by a federal official to be "the safeguard and life of the Territory."[25]

The USRC *Lincoln*, anchored at the port of Victoria in British Columbia, was the first cutter to cruise Alaskan waters after the U.S. purchase of the territory in 1867. However, in the last year of the Civil War (1865), Captain C.M. Scammon (USRM) sailed the RC *Shubrick* to Russian America. Captain Scammon's mission was in support of the failed attempt by the Western Union telegraph company to lay a transmission line across the formidable Bering Strait, and from Russian Siberia to European Russia, to the port of St. Petersburg. The *Lincoln* brought Lt. George W. Moore (USRM) to represent the United States in Sitka in 1867. From 1867 to 1869, Capt. J.W. White (USRCS), in the RC *Wayanda*, surveyed coastal Alaska from the Aleutians north through the Bering Strait. "Other cutters from West Coast ports that became known to Alaskan natives, traders, and immigrants in that early period were the *Wolcott*, *Rush*, and *Corwin*."[26] U.S. Navy authorities, along with U.S. Marines, the collector of customs, and policy coordinators in the U.S. War Department and U.S. Treasury, crafted effective policies and curbed criminality and Indian disturbances. Cutters and crews supported and transported political, judicial, and law enforcement authorities in the settled communities and served as the only United States officials who administered law and order and federal policies in the remote maritime and coastal regions. The official Bering Sea (and Aleutian Islands) Patrol by a half-dozen revenue cutters under a force commander began in the 1890s, although in the previous decade a cutter had reached Point Barrow on the north coast of Alaska. The revenue cutter missions would involve coastal survey, search and rescue, law enforcement, scientific exploration, charting and cartography, and federal investigations.

The USRCS patrols of the 165-foot RC *Lincoln*, after 1867, were commanded by the very able Capt. W.A. Howard, along with his skilled and courageous crews and the highly competent Lt. George W. Moore. The *Lincoln* carried a U.S. Coast Survey team led by George Davidson and a medical officer.

The legendary RC *Bear* added immeasurably to the maritime missions and legacy of the USRCS and USCG. The RC *Corwin* (Capt. J.W. White) brought professionalism, service, and positive notoriety to the USRCS in the foggy, icy waters of the region, as would the cutters *Forward* and *Dallas* and the courageous crews of the U.S. Lighthouse Service and surfboat sailors of the U.S. Life-Saving Service in the last decades of the 19th century and after. Fishing and seal hunting enforcement became essential law enforcement specialties of the USRCS in the forbidding weather, waters, ice, and fog of the high latitudes. The U.S. Coast Survey scientists gathered statistics and other data for the construction of maps and charts in those waters, and revenue cutter crews, including the scientifically inclined surgeons, secured flora and fauna data and specimens for the scientists at the Smithsonian Institution[27] in Washington, D.C.

The cutter missions in Alaskan waters were as complex and dangerous as they were essential and creative. Surveys were made for coaling station locations to fuel the cruising patrol craft of the Revenue Marine. Smuggler haunts, rich fishing banks, and coastal physical geography were explored, and, as in the case of the RC *Wayanda*, exploring parties were sent up various rivers to seek out settled communities, farmland, natural resources such as coal, and good harbor locations. Codfish and halibut schools and undersea topographic banks that sheltered and nourished marine life were discovered and mapped. The captains and crews did their share of extraordinarily successful fishing. The cutter teams worked with the U.S. Bureau of Fisheries to establish procedures and regulations that enhanced productivity, good conservation practices, and effective law enforcement to conserve fish and seals. Sealing companies were informed, educated, and monitored. Poachers and other law violators were apprehended and fined. Soundings, charts, and hydrographic and oceanographic observations were made and recorded. Fishing and hunting regulations were established.[28] Other cutter fleets and crews were sent north in 1895, initially in the Bering Sea Patrol Force under Captain C.L. Hooper (USRM), to perform— as Professor David S. Jordan, a member of the Seal Commission described it— "work with the greatest faithfulness by the admirably organized fleet of United States revenue cutters, in unpleasant, dangerous, continuous cruising in rough and foggy seas, while [the missions] of examination and seizure [were] extremely unpleasant."[29]

In 1880 the USRC *Corwin*, under the command of Capt. C.L. Hooper, commenced Arctic Ocean cruising patrols north of 66 degrees north latitude. Those high-latitude missions included SAR, assistance to ships and crews, law enforcement, the support of the U.S. Army Signal Corps mission, and communication with and care of settlers, seamen, and Native Americans along the U.S.

Arctic shoreline. In summation, the USRC missions, as Coast Guard historian Captain Stephen H. Evans concluded, were dedicated to providing navigation access, maritime rescue, and transportation and safety to the inhabitants, seaways and harbors of Alaska. Captain Hooper's first mission in 1880 and 1881 was dedicated to finding the lost exploration steamer USS *Jeannette*, under the command of Lt. Cmdr. G.W. De Long (USN), and two missing whaling vessels, SS *Mount Wollaston* and SS *Vigilant*. The search involved traversing the waters of the Bering Sea and placing a landing party with sled dogs in extreme weather along the Asian coast. The loss of the rudder on the RC *Corwin* could have doomed the crew complement had not the skilled cutter carpenter created a substitute. The whalers were not found, but wreck remnants and stories from Inuit hunters indicated the crews had become iced-in and starved to death. A wandering remnant of crew survivors would later reveal the fate of the USS *Jeannette* and her helpless drift to destruction in the solid ice pack that imprisoned the vessel.[30] Stephen H. Evans described the hazardous Arctic challenges Captain Hooper and his crews navigated with riveting detail, as "the *Corwin* was once more ready to steer handily through the fogs and floes."[31]

The contributions the USRCS and the contemporary USCG have made using their cutters as scientific platforms to study oceanography and the Arctic are well illustrated by the early voyages of the RC *Corwin* under the command of Captain Hooper. Among the scientists who sailed on the *Corwin* was the famed naturalist John Muir, who then wrote about Arctic and subarctic glacial origins and activities and how the land bridge between Siberia and North America, which supported the migration of America's first migrants, was caused by warming climates and the glacial erosion that exposed the intercontinental pathway across the Bering Sea and down the Aleutian chain. Another famous scientific passenger on the *Corwin* was the naturalist E.W. Nelson, whose writings on what would later be called ethnic anthropology were enlightening. Nelson's parallel natural history collections in Alaska supplemented Professor Muir's botanical studies.

Captain Hooper and his well-educated officers, including ship's physician Dr. Irving G. Rosse, took good care of the crew complement and hundreds of needy native Inuit. Rosse's published medical and anthropological notes about the appreciative Native Americans were valuable scientific and policy-making contributions. Captain Hooper added to the scientific and geographic literature of the high latitudes with his own complete and perceptive cruise reports and log entries. The cutter commander was concerned about the lives and health of the hundreds of indigenous Americans with whom he and his crew came in contact, learned from, and assisted.[32] The Revenue Cutter Service voyages and scientific contributions led Congress to craft more effective humanitarian policies for the administration of the native populations—first undertaken under the auspices of the U.S. Bureau of Education and then the U.S. Office of Indian Affairs—that the cutters supported "by transporting agents, wards, equipment, and supplies throughout the length and breadth of the northern littoral."[33] Captain Hooper's concern for the Native Americans was reflected in the way he dispensed "fair dealing, swift justice, and strict enforcement of liquor laws," including the interdiction of contraband from American schooners. But Hooper did not enforce federal regulations prohibiting the sale of rifles to the natives (ostensibly to protect white settlers from the tribes), because he believed the weapons would be useful in hunting game and providing essential food for the "aboriginal populations" in marginal resource areas.[34] Captain Hooper's sensitivity and concern for the Native Americans led him to advise that the Arctic Alaskans not be called Eskimo but addressed by the name they used for themselves: Inuit. The term "Eskimo" was a derogatory term meaning "raw fish eaters"[35] (which in modern times is in fact a sophisticated culinary concept).

The Revenue Cutter Service expanded its patrols and missions under Captain Hooper, who made his headquarters in the town of Unalaska at the port of Dutch Harbor in the Aleutian Islands. From there, revenue cutters enforced sealing laws and seized U.S. and foreign fishing and hunting vessels that violated federal laws. Periodic international agreements between the United States and the seal hunting nations of Japan, Canada, and Russia ameliorated often-tense and dangerous confrontations between the USRCS and contraband vessels and crews.[36] In SAR missions for lost vessels and missing sailors, Captain Hooper instructed his courageous landing party crews, under the able command of First Lt. William J. Herring, assisted by Inuit guides, drivers and interpreters, to search mainland Asian-Siberian territory because marooned seamen would likely seek safety on shore. The cutter landing parties would carry several months of food provisions, sleds and dogs, and light skin boats to facilitate open-water crossings.[37]

The Revenue Cutter Service established and supported a hospital at Unalaska that served fishermen, Native Americans, the Russian ethnic population, merchant seamen, whalers, sealers, and U.S. Navy and Revenue Cutter personnel. Captain Hooper's forwarding of physician and other medical staff reports to the Treasury secretary facilitated the eventual funding and staffing of a permanent Marine Hospital Service. The support and funding was sporadic, but another creative and dedicated Bering Sea commander, Captain A.J. Henderson, facilitated the eventual full staffing and provisioning of the hospital that would serve the cuttermen.[38] The RC *Thetis*, like other cutters in the Bering Sea Squadron, established an exemplary reputation patrolling the Aleutian

Islands to police seal poachers, enforce customs laws, do SAR missions, apprehend fugitives from justice, administer law and order, apprehend whiskey smugglers, tow disabled whaling ships to harbors, administer to ill and hungry people, support scientific expeditions, transport federal judges and hold floating court sessions onboard, and provide ferry service across the waterways. After decades of hard Arctic sailing and service, the USRC *Thetis* was decommissioned in 1916.[39]

We have already considered the roles that President George Washington, the U.S. Congress, and Treasury Secretary Alexander Hamilton played in creating the U.S. Revenue Marine to serve as the tariff and customs duties collector and first naval arm of the new Federal government. The Revenue Marine teamed up with the fledgling U.S. Navy in the Quasi-War with France (1800) and the Second War of Independence against Britain in the War of 1812. The maritime skills of the Revenue Marine challenged the British in the coastal and interior waters of the United States, forcing the United Kingdom to split its land and sea forces over extensive land areas, thus diffusing the strength of the world's premier military and naval power and establishing the substantive mutual alliance system that benefited and enhanced the powers of both maritime nations.

Alexander Hamilton gained his military experience in the Revolutionary War as a commanding officer of American troops and as the indispensible military aide of General George Washington. Hamilton learned military and naval strategy from General Washington and from his own experiences in ways that helped him administer the Revenue Marine in its multi-mission domestic and national defense responsibilities while secretary of the Treasury in the years 1789–1795. That military experience no doubt contributed to Hamilton's effective application of naval training and mission strategies in his administration of the U.S. Revenue Marine, the future United States Coast Guard.

In his writings, naval historian Lt. Cmdr. Thomas J. Cutler, USN (Ret.), credited Gen. Washington with having the logistical and tactical skills worthy of both a general and an admiral. The French navy was instrumental in helping the colonial Americans achieve victory over the British, an alliance that Washington fostered. Lieutenant Commander Cutler credited General Washington with perceptively adjusting to overwhelming British power by modifying his troop movements to take advantage of America's extensive land areas and coastal and interior waters and peninsulas, and splitting British naval, marine and army forces and separating them from each other. Those tactics were instrumental in causing the British defeat and surrender at Yorktown, Virginia,[40] and the subsequent Treaty of Paris (1783) that ended the war and gained freedom and sovereignty for the United States.

The Revenue Marine missions in the Bering Sea, the waters off Alaska, and in the Arctic polar region established the historic and contemporary presence of the United States Coast Guard beginning in 1867. That presence became especially significant in the early 21st century when Arctic ice shifts exposed strategic waterways that attracted oil and mineral exploitation firms, commercial shipping, passenger vessels, and international territorial claims and illuminated the economic and national defense interests and presence of the Polar Rim nations of Scandinavia, Canada, Russia, and the United States. The Coast Guard would become the lead agency in guarding against and responding to environmental pollution in the maritime realm of the United States, as well. Commercial ship operations would require an increased presence of Coast Guard assets and infrastructure in the far North for SAR, navigation and law enforcement missions. The increased presence of commercial shipping and naval vessels would also require enhanced vessel hull construction, ship safety drills, and training for the variable weather and ice conditions of the high latitude regions, and the clothing and equipment that could withstand those conditions.

The strategic interests of the United States stimulated the movement to construct and procure new, expensive, state-of-the art United States icebreakers to compete numerically and technologically with the advanced icebreakers of the other Arctic nations.[41] The U.S. Coast Guard would be building on the precedents established by the Revenue Cutter Service of the late 19th and early 20th centuries in the high latitudes, and the early icebreaking and multi-mission services of the reinforced, wooden-hulled cutters of that day.

14

The Spanish-American
War (1898)

Following the initial naval and military policing of the Bering Sea and Alaskan terrestrial and maritime frontiers, the immediate national defense conflict that challenged the U.S. Revenue Cutter Service, U.S. Navy, U.S. Marines, and U.S. Army was the war between the imperial monarchy of Spain and the United States. The geostrategic realm of the conflict included Spanish Cuba in Latin America and the Caribbean Sea and the Spanish Philippines in the far western Pacific Ocean. When the war ended, the United States had absorbed those Spanish colonies. By definition, and after having somewhat hypocritically criticized Spanish colonialism and imperialism, the United States stepped onto the international stage as a fledgling colonial power, rivaling the territorial realms of the Netherlands, Britain, France, Germany, and Belgium. The United States then found itself involved in international power struggles, wars, and geopolitical responsibilities to the present day. The resultant cost-benefit ratio has been controversial and costly.

The 149-foot USRC *Winona* sailed into the Gulf of Mexico to protect the southern coast of the United States during the Spanish-American War of 1898. The 219-foot steam-powered sailing barque, the USRC *McCulloch*, served in the squadron of Commodore George Dewey (USN) at Manila Bay in the Philippines during the war. As a U.S. Navy auxiliary vessel, the combat cutter featured a bow torpedo tube and four three-inch deck guns. By the time of the Spanish-American War, the U.S. Revenue Marine was being referred to as the U.S. Revenue Cutter Service (USRCS) in official federal documents and communications. During the war the USRCS engaged in joint missions with the U.S. Navy in Cuba in the maritime areas of Havana Harbor and off the port of Cardenas. Off Cardenas, the USRC *Hudson*, under the command of Lt. Frank Newcomb (USRCS), towed the damaged navy torpedo boat USS *Winslow* to safety under heavy Spanish coastal and naval gunfire. Captain Newcomb would later receive the Congressional Gold Medal for his courage, competence and tactical combat decision. Worth G. Ross entered the U.S. Revenue Cutter School of Instruction (the future U.S. Coast Guard Academy) as a cadet in 1877, in the first graduating class of that officer training institution. He later served on the USRC *Woodbury* and USRC *Harvard* and received a Bronze Medal from Congress for his combat actions off Santiago, Cuba, in the Spanish-American War. Captain Ross would become the captain-commandant of the U.S. Revenue Cutter Service in 1905.[1]

The causal factors of the Spanish-American War are complex. But the chronological period between the end of the Civil War and the Spanish-American War, covering the decades between 1865 and 1898, were challenging times for United States military and naval forces. The U.S. military was busy policing the migration of Americans across the Trans-Mississippi West and the Indian frontiers, as well as attempting to bring stability and law and order to the post–Civil War Reconstruction South between 1865 and 1877. The post–Civil War duties of the U.S. military required increased budgets, training and personnel. The Euro-American migration across the Trans-Mississippi West led to the involvement of the United States military in the hard-fought Indian Wars, which continued until the closing of the frontier in 1890. The colonial result of the Spanish-American War of 1898 required expanded American military and naval forces to subdue indigenous resistance in the territories acquired from Spain by America, which included Cuba, Puerto Rico, and the Philippines.

Between 1898 and the 1917 entry of the United States into World War I, American naval and military personnel would suppress conflicts and establish control in the Caribbean region, invade Mexico, suppress the Filipino insurrection, and support America's new imperial and economic interests in Asia and the Pacific. The United States was drawn into the war with Spain ostensibly to end the evils of Spanish colonialism, but it ended the war as a colonial power. From those geostrategic interests, the United States would compete with Russia, Japan, France, Britain and Germany, join forces to crush nationalistic rebellion in China, and become involved in World War I, World War II, and Asian and Middle East wars in the 21st century.[2]

To begin the history of American expansion beyond the North American continent, the origins of the Spanish-American War must be considered. Americans did have sympathy for the Hispanic peoples suffering under the yoke of Spanish colonialism, in part because of America's fight for independence from British control of the American colonies. The American press sympathized with the Cuban nationalist movement against Spanish control, and generated

American hostility toward Spain with true and false stories of Spanish oppression. The exaggerated, biased, and warmongering stories about Spanish oppression, some of it fueled by anti–Catholicism, sold newspapers and would be labeled "yellow journalism."

The sinking of the visiting battleship USS *Maine* in Havana Harbor on 18 February 1898 with the loss of more than 200 U.S. Navy personnel fueled the flames of war because the explosion that sank the ship was attributed to Spanish sabotage (after the war, experts concluded that the explosion was internal to the ship and likely caused by an accidental explosion of coal dust and vapors that collected in the ship's fuel bunkers). Public opinion, national pride, and political pressure prompted the United States government to formally declare war on Spain on 25 April 1898. The U.S. Navy was immediately ordered to send warships to engage Spanish military and naval forces in Cuba in the Caribbean Sea and in the Spanish Philippines in the eastern Pacific Ocean.[3] Robert Pendleton and Patrick McSherry have chronicled the missions of the U.S. Revenue Cutter Service on the Spanish-American War Centennial website.[4] The authors reviewed the cutters involved within and outside the theaters

The RC *Hudson* in 1898 at Norfolk Navy Yard before its combat action in Cuba.

The 219-foot RC *McCulloch* (Capt. D.B. Hodgsdon) entered Manila Bay in the Philippines (1898) under the command of Commodore George Dewey (USN). The cutter performed battle station and dispatch duties for the U.S. Navy Squadron. The Spanish fleet was destroyed in the battle.

of war, the characteristics of the cutters, the crew complements, and the missions and achievements of particular revenue marine warships.

After the U.S. declaration of war on Spain on 21 April 1898, President McKinley ordered the temporary transfer of the USRCS to the authority of the U.S. Navy secretary and directed that the U.S. Revenue Marine cutters become part of the Auxiliary Naval Force (ANF). The ANF would be composed of personnel and vessels from state naval militias, the U.S. Fish Commission, the U.S. Coast Survey, and the predecessor agencies of what would become the

Opposite, top: The 149-foot USRC *Winona* patrolled the Gulf of Mexico out of Mobile, Alabama, in the Spanish-American War (United States Coast Guard). *Bottom:* The sinking of the 394-foot battleship USS *Maine* in Havana Harbor, Cuba, during the Cuban revolt stimulated anti–Spanish war fever in America. Three-fourths of the 374 USN crew complement were killed in the internal explosion, initially thought to be caused by a Spanish mine (Naval Historical Foundation).

U.S. Coast Guard, in 1915, including the U.S. Lighthouse Service[5] and personnel and craft as might be needed from the U.S. Life-Saving Service. Thirteen Revenue Service cutters coordinated with the U.S. Navy, eight with the North Atlantic Squadron, and one, the RC *McCulloch* (Captain Daniel B. Hodgson, USRCS), with 10 officers, 95 enlisted men, and 6 deck guns, was assigned to Admiral George Dewey (USN) and his Asiatic Squadron. Four cutters (*Corwin, Perry, Rush,* and *Grant*) patrolled the Pacific waters from California to Alaska to protect commercial vessels from enemy warships and privateers. The four cutters housed 30 officers and 128 enlisted personnel and carried 12 guns. The North Atlantic Squadron contained 339 enlisted personnel, 58 officers, and ordnance of 43 deck guns. The entire USRCS had 13 operational cutters with a crew complement of 98 officers and 562 enlisted men and carried 61 guns.[6] The RC *McCulloch* joined Adm. Dewey's U.S. Navy Squadron out of British Hong Kong in China and proceeded as a guard ship for U.S. Navy supply transports to combat action at Manila Bay in the Philippines. It also served as a messenger vessel for Adm. Dewey from the Philippines to Hong Kong. The crew of the *McCulloch* would capture the *Leyte,* a gunboat of the Spanish navy.

The U.S. Navy blockade of Spanish Cuba in the Caribbean Sea consisted of U.S. Navy warships and eight U.S. Revenue Service cutters assigned to the U.S. Navy Squadron in those waters, under the command of Rear Adm. William T. Sampson (USN). On May 11 1898 the RC *Hudson* (Lt. Frank Hamilton Newcomb, USRCS) rescued the combat-damaged U.S. Navy torpedo boat USS *Winslow,* towing it out of harm's way at the Battle of Cardenas under heavy Spanish artillery fire.[7] After the *Hudson* rescue, a postwar federal document would conclude "the Revenue-Cutter Service is prepared today to act as an auxiliary to the Navy in any scheme of defense. Its vessels are armed and its officers and men are organized, instructed, and drilled very much as in the naval service. In future wars, this Service will perform its proper duties—those of a coast-guard navy—and be supplemented by an auxiliary coast defense fleet."[8]

The shipbuilder David Bell of Buffalo, New York, built the USRC *Hamilton* in 1870–1871 and named it for President George Washington's Treasury secretary, who administered the U.S. Revenue Marine after its 1790 creation by Congress. The iron-hulled vessel carried one 4-inch gun and a complement of 38 men. The *Hamilton* served with the North Atlantic Squadron and joined the naval blockade off Havana, Cuba, in the war. It delivered mail and experienced enemy gunfire, then sailed out of Charleston, South Carolina, after its mission with the U.S. Navy ended. The *Hamilton* officer complement included Captain W.D. Roath, four lieutenant grade officers, three engineers, and the medical officer, ship surgeon Dr. Charles H. James, Jr.[9]

The Globe Iron Works in Cleveland, Ohio, completed the construction of the USRC *Algonquin* in 1897, one year before the start of the war. The 205-foot cutter listed a crew roster of 71 sailors and was capable of a speed of 16 knots. The two-gun vessel served in the North Atlantic Squadron. First Lt. W.C. DeHart was the commanding officer. Lt. DeHart's officer colleagues included two second lieutenants, one chief engineer, a first assistant engineer, and a second assistant engineer. The rates of the enlisted crew members provide an appreciation of the complex duties and specialized skills needed to run a late 19th-century steam-powered revenue cutter. The rates and specialties of the crew members included stewards, ordinary seamen, firemen, gunners, small boat coxswains, seamen, cooks, boatswains, oilers, carpenters, a master-at-arms, coal passers, and machinists.[10]

The U.S. Revenue Cutter *Boutwell* sailed out of the port of New Bern, North Carolina. The USRC *Calumet* was commanded by First Lt. W.H. Cushing. The two-gun cutter sailed with the U.S. Navy in the North Atlantic Squadron. Lt. Cushing's commissioned colleagues included W.G. Blasdell, the executive officer who was second in command; First Asst. Engineer A.J. Howison; and Second Asst. Engineer Urban Harvey. The RC *Chippewa* was assigned to patrol the Great Lakes during the Spanish-American conflict. The 165-foot RC *Commodore Perry*, commissioned in 1884, was an auxiliary (sail- and steam-powered) ship constructed by the Union Drydock Company in Buffalo, New York. The cutter was iron-hulled and propelled by sails and a single propeller. The *Perry* was assigned to the Pacific Squadron with the mission of protecting U.S. commercial vessels and gold-bearing ships sailing between the Klondike region of Canada and Alaska and San Francisco, California. The Klondike and Yukon River regions of western Canada and the area around Nome, Alaska, were the sites of gold rush miners and civilians between 1896 and 1899. First and second lieutenants, two third lieutenants, a chief petty officer, two second assistant engineers, and a ship surgeon, Dr. W.L. Ludlow, supported Captain W.F. Kilgore (USRCS), the *Perry*'s commanding officer.

The RC *Corwin*, a wooden-hulled vessel, sailed with the Pacific Squadron and plied the waves between the gold fields of the Klondike region and San Francisco. A first lieutenant who served as the executive officer, three second lieutenants, one of whom was the navigator, three engineering officers, and surgeon Stephen Whyte assisted Captain W.J. Herring, the commanding officer of the *Corwin*.[11] The 185-foot USRC *Dexter* patrolled out of Narraganset Bay, Massachusetts. The RC *Galveston*, constructed in 1891 by Reeder & Sons of Baltimore, Maryland, defended the lower Mississippi River and the port of New Orleans, Louisiana, with a complement of 58 crew members. The 205-foot RC *Gresham*, built in 1896 by Globe Iron Works in Cleveland, Ohio, could do more than 15 knots and had a crew complement of 103 officers and men.

The steel-hulled, single-screw cutter guarded the coastal region of the northeastern United States as part of the North Atlantic Squadron and supported U.S. Navy missions with its four rapid-fire deck guns.[12]

The USRC *Hudson*, mentioned previously, was a significant naval warship in the war despite its small size. Built by the Camden, New Jersey, John H. Dialogue Shipyard and commissioned in 1893, the 94-foot cutter had a 10-foot draft, could do 12 knots, and carried a 23-crew complement and three guns: two 6-pounders and a Colt automatic machine gun. A 525-horsepower engine propelled the steel-hulled vessel. Lt. F.H. Newcomb (USRCS) commanded the warship in the North Atlantic Squadron of the U.S. Navy. Off the Cuban coast and the port of Cardenas the RC *Hudson* towed the battle-damaged torpedo boat USS *Winslow* away from a devastating Spanish artillery barrage and interdicted two Spanish fishing sloops that tried to penetrate the U.S. Navy blockade of Spanish Cuba. The *Hudson* delivered official dispatches and mail to various sites and vessels. The officer complement of the RC *Hudson* included First Lt. Newcomb, Second Lt. (and executive officer) J.F. Scott, Third Lt. E.E. Mead, and first and second assistant engineers N.E. Cutchin and Ted Lewton. The specialized enlisted complement in the battle off Cardenas included coal passers, firemen, seamen, a cook, steward, boatswain, and two quartermasters,[13] the latter specialization a naval rate that involved navigation, signal, and helmsman duties on the bridge. E.F. Johnson, the cook on the *Hudson*, was, in the parlance of that historical period, a Negro.[14]

The Revenue Cutter *McCulloch* was a significant USRC warship, as previously noted. Built by William Cramp Shipyards in Philadelphia, Pennsylvania, the 219-ft. cutter had a 33-ft. beam and a 14-ft. draft. The vessel could sail at 17 knots and carried a crew complement of 68 officers and men. The *McCulloch* served with the U.S. Navy in the Asiatic Squadron and at the Battle of Manila Bay in the Spanish Philippines.[15]

In 1881, H.A. Ramsey Shipyards in Baltimore, Maryland, completed the construction of the RC *Guthrie*, which guarded the waters around Baltimore during the war. J.W. Lynn and Sons in Philadelphia constructed the 138-ft. USRC *Woodbury* in 1863 and 1864. The wooden-hulled vessel was powered by a single screw, carried seven deck guns and had a crew complement of 41. The cutter sailed with the U.S. Navy in the North Atlantic Squadron and participated in the naval blockade of Havana. The wartime crew roster included Captain H.B. Rogers and First Lt. W.G. Ross, the executive officer.[16] In his subsequent career, Captain-Commandant Ross would assume command of the U.S. Revenue Cutter Service, in 1905. During the Spanish-American War, First Lt. Ross served on the Navy vessel USS *Harvard* and earned the Bronze Medal from Congress for his actions in naval combat off Santiago, Cuba.[17] The officer

roster included two second lieutenants, one third lieutenant, a chief petty officer, one first and one second assistant engineer, and Dr. Edward F. McConnell,
the ship surgeon. The enlisted complement included the rates of steward, cook,
oiler, gunner's mate, fireman, seamen, coxswain, ordinary seaman, senior master, coal passer, and carpenter.[18] The enlisted crew included five Negro sailors.[19]

First Lieutenant William E. Reynolds (USRCS) commanded the RC
McLane in the Spanish-American War. The *McLane* was built in Wilmington,
Delaware, as a side-wheel steamer. Named the RMS (Revenue Marine Service)
Louis McLane in 1873 the cutter was designated the USRC *McLane* in 1894. The
iron-hulled cutter was assigned to the North Atlantic Squadron. The cutter's
officer roster included two second lieutenants, one of whom served as the executive officer, a third lieutenant, and a second assistant engineer.[20] Lt. Reynolds
would go on to serve as commandant of the U.S. Coast Guard from 1919 to
1924, from just after the end of The Great War (World War I) to the early days
of the controversial years of Prohibition enforcement. In 1923 Reynolds
became the first commandant to be promoted to the rank of Rear Admiral.[21]

The 205-ft. USRC *Manning* was constructed by the Atlantic Works at
Boston, Massachusetts, and was commissioned in January 1898. The single-
screw, composite-hulled vessel boasted a 2,000-horsepower engine, three deck
guns, and a complement of 10 officers and 65 enlisted personnel. The *Manning*
had an officer roster that included the commanding officer, Capt. F.M. Munger,
a first lieutenant who was the executive officer, one second lieutenant, three
third lieutenants, a chief engineer, one first assistant and two second assistant
engineers, and ship surgeon Dr. A.T. Mitchell. Among the officer responsibilities were ordnance and navigation.[22] Second Lieutenant Godrey L. Carden
(USRCS), the ordnance officer of the RC *Manning* responsible for training and
administering the gun crews, weapons, and ammunition, would go on to serve
as the U.S. Coast Guard captain of the port (COTP) in New York City, through
the U.S. involvement in World War I (1917–1918).[23]

Captain H.D. Smith (USRCS) commanded the 145-ft. RC *Morrill*. The
cutter was constructed at the Pusey and Jones Shipbuilding Company in Wilmington, Delaware, in 1889, nearly a decade before the outbreak of the Spanish-
American War. The iron-hull, single-propeller cutter carried two three-pounder
deck guns on patrols with the U.S. Navy in the North Atlantic Squadron. Captain Smith's nine-officer complement included first, second and third lieutenants, and a chief engineer, a first assistant engineer, and a second assistant
engineer. Dr. J. Spencer Hough was the ship surgeon.[24]

Globe Iron Works in Cleveland, Ohio, constructed the 205-ft. RC *Onondaga*. The cutter displaced 1,190 tons, could travel at 16 knots, and carried a
complement of 73 men. The RC *Rush* was built in Boston, Massachusetts, at

the Atlantic Works Shipyard. Commissioned in 1873 with a crew of 40 men, it served in the Spanish-American War on the U.S. Navy Pacific Squadron, protecting commercial vessels that sailed the waters between the Klondike Canada-Alaska gold fields during the war. The cutter carried three deck guns and had an officer roster of nine, including Captain W.H. Roberts (USRCS), one first and one second lieutenant, two third lieutenants, a chief engineer, two second assistant engineers, and a physician, Robert McAdory. Constructed by Pusey and Jones at Wilmington, Delaware, in 1870–1871, the RC *Ulysses S. Grant*, named after the Union Civil War general and U.S. president (1869–1877), served with the U.S. Navy's Pacific Squadron protecting ships sailing between San Francisco and the gold fields of Canada and Alaska. The iron-hull cutter was propelled by a single-screw propeller. Captain J. A. Slamm (USRCS) had four officers under his command, including an executive officer, navigator, surgeon (Dr. Robert Hammond), first, second and third lieutenants, a chief engineer, and two second assistant engineers. The 170-ft. RC *Windom* was partly constructed by Iowa Iron Works in Dubuque, Iowa, completed in Baltimore, Maryland, and commissioned in 1896. The single-screw, steel-hull, 800-horsepower cutter carried one gun in her service with the U.S. Navy's North Atlantic Squadron. The commanding officer of the *Windom* was Captain S.E. Maguire, whose 9 fellow officers included a first lieutenant executive officer, two second lieutenants, one third lieutenant, a chief engineer, a first and second assistant engineer, and two—consecutively serving—surgeons, Dr. John C. Travis and Dr. W.E. Handy.

Pusey and Jones Shipbuilding Company in Wilmington, Delaware, constructed the 149-ft. USRC *Winona*, which was commissioned in 1890, eight years before the Spanish-American War erupted. The *Winona* patrolled out of the port of Mobile, Alabama,[25] during the war to protect American commercial vessels and ports and the maritime domain between Cuba and the United States. Captain Charles F. Shoemaker was the USRCS commandant during the Spanish-American War and served as head of the Revenue Cutter Service from 1895 to 1905. In 1895 U.S. Treasury secretary John G. Carlisle ordered the Revenue Cutters *Winona*, *Forward*, and *Morrill* to patrol with the RC *McClane* around the Florida peninsula and north to Wilmington, North Carolina, to enforce U.S. neutrality and health quarantine laws. Attempts by American sympathizers and Cuban rebels fighting against Spanish colonial military and naval forces to smuggle arms and personnel had to be stopped in the years before the United States entered the war against Spain. The *Winona*, commanded by Capt. Charles A. Abbey, seized the contraband schooner *Lark* at the Florida port of Bahia Honda. In December the crews of the Revenue Cutters *Forward* and *Winona* captured a rebel camp at Cape Sable, Florida, that housed medical and military supplies and had training facilities estimated to support 50 people. After the entry of the United States into the Spanish-American War, the *Winona* joined other cutters to enforce the Cuban blockade, escort U.S. Army transport ships to battle areas, and protect U.S. vessels from Spanish mines.[26]

Captain Stephen H. Evans (USCG) considered the Spanish-American War and the role of the USRCS in it in significant detail. The narrative Evans crafted contained voluminous primary source quotes from federal documents and correspondence.[27] He credited contributions of the USRCS to the U.S. Navy mission as having "revived public awareness of the Revenue-Marine's military function, and prepared the way for the unequivocal statutory definition of the cutter establishment's military status,"[28] in addition to its traditional congressionally and federally mandated civil duties, and giving officers and enlisted personnel rank and rate pay and retirement benefits that would be comparative to the U.S. Army, Navy, and Marine Corps.[29] Evans covered revenue cutter missions in the prewar period of 1895–1898, with the Florida Straits patrols of the revenue cutters *Morrill, Colfax, Forward, Boutwell, McLane,* and *Winona* enforcing neutrality laws and holding suspected ships in port. Assistant U.S. Navy Secretary and future president Theodore Roosevelt ordered the ships of Commodore George Dewey (USN) to Hong Kong to prepare for what became the U.S. invasion and absorption of the Spanish Philippines. Roosevelt's 25 February 1898 instructions to Dewey were "to see that the Spanish squadron does leave the Asiatic coast, and then do offensive operations in Philippine Islands."[30]

The North Atlantic Squadron under the command of Commodore (later Rear Admiral) William T. Sampson (USN) was ordered to combat and defeat the Spanish in Cuba and protect U.S. coastal towns and American commercial vessels. The USRCS was ordered to assist the navy with its cutters and serve as scouting and dispatch boats, gunships, and transports. In a 9 March 1898 headline The *San Francisco Examiner* stated, "More Vessels Wanted [for the U.S. Navy]." The article explained that Revenue cutters and Lighthouse tenders would be assigned to the navy. A later *Examiner* headline read, "Revenue Cutters to Be Pressed into Service of the Navy," and a 24 March 1898 article reported, "The Navy Department today took steps toward utilizing ten seagoing revenue cutters. Captain Shoemaker, Chief of the Revenue Cutter Service, conferred with Assistant Secretary Roosevelt on plans for turning over these cutters to the Navy. They will be first sent to Norfolk [Virginia] where additional guns will be mounted and then proceed to Key West to become a part of the squadron there, and serve as naval pickets ... outside [flanking] the [U.S. Navy] cruisers and battleships."[31] The famed Revenue Cutter *Bear* and other cutters would be armed, and the Berlng Sea Fleet missions would be terminated so those cutters could be dispatched to protect the Pacific Coast from potential

Spanish raiders and ease the fears associated with West Coast ports officials and residents.

The USRCS prepared for service approximately "24 line officers, 74 engineering officers, 900 enlisted men, and 19 vessels [including the Revenue Cutters] *Manning, Gresham, Algonquin, Onondaga,* and *Windom,* all new, fast and efficient vessels that can go anywhere and perform any service that any vessel of their class can perform."[32] The *Examiner* added, on April 15, "It is an open secret that the cutters *Rush, Corwin,* and *Grant,* now in port, will not go north this season. The cutters *Bear* and *Golden Gate* are also on the [Pacific] coast."[33] Captain G.H. Gooding, on the RC *Winona,* conducted a raid on enemy commerce. A dispatch from the *San Francisco Examiner* from Biloxi, Mississippi, headlined the event—"Revenue Cutter *Winona* Takes Spanish Steamer"—and added that the capture occurred at Ship Island, on the Mississippi River.[34] Two weeks before the capture of that Spanish steamship, Assistant Treasury Secretary W.B. Howell had informed officials by cablegram from Washington, D.C., that the American counsel-general was to direct Captain D.B. Hodgson of the revenue cutter, now the USS *McCulloch,* to "proceed to Hongkong [*sic*] [for duty with] Commodore Dewey at the Asiatic Station."[35]

Upon arrival at Commodore Dewey's command, the *McCulloch'*s bunkers were filled with coal, the hull painted the Navy gray color, and naval signal books and night signals, squadron orders, and the cipher code were given to Lt. John Mel (USRCS), who was assigned to the position of signal officer. Moving out with Dewey's squadron on 27 April, the *McCulloch* took its naval convoy position on the outer right flank to protect the navy supply ships USS *Nanshan* and *Zafiro.* The USN cruisers and gunboats in the squadron were the USS *Boston, Raleigh, Baltimore, Petrel* and *Olympia.* On Revenue Marine and U.S. Navy vessels, crews stowed gear, performed battle station drills, and prepared for naval warfare. On 30 April the naval squadron arrived at what would later become the U.S. Navy Base at Subic Bay. Just after midnight on 1 May 1898 Spanish gunners fired at the squadron. Naval vessels fired back. The *McCulloch* fired three shells from her aft-starboard gun, then deployed to protect the USN supply ships and tow them from danger should the need arise. By noon the battle was ended.[36] The *Examiner* reported the victory: "Immediately after the engagement, Adm. Dewey sent the Revenue Cutter *McCulloch* back to Hongkong [*sic*] with dispatches."[37]

The RC *Hudson'*s rescue of the damaged USS *Winslow* at Cardenas Bay in Cuba won acclaim from the press and from President McKinley, who wrote a letter to Congress from the Executive Mansion dated 27 June 1898, in which he cited the 11th of May naval conflict in the bay. Among the president's observations and accolades were the following statements: "In the face of a most galling fire from the enemy's guns, the revenue cutter *Hudson*, commanded by First Lieutenant Frank H. Newcomb, United States Revenue Cutter Service, rescued the disabled *Winslow*, her wounded commander, and remaining crew. The commander kept his vessel in the very hottest fire until he got a line made fast to the *Winslow* and towed that vessel out of range of the enemy's guns."[38] President McKinley recommended that Congress present a gold medal to Capt. Newcomb, silver medals to the officers, and bronze medals to the crewmen. The United States Congress followed the president's recommendation by joint resolution.[39]

The cutters *Woodbury, McLane, Morrill, Hamilton*, and *Windom* (under Capt. S.E. Maguire) escorted a damaged U.S. Navy vessel out of gunfire range. All the cutters and officers and crews performed courageously under enemy guns close ashore and destroyed several Spanish stone fortifications.[40] The USCGC *Manning* (Captain Fred M. Munger) experienced a variety of missions in the Cuban war. Captain Munger submitted a history of the *Manning*'s actions, and the professional responses of the crew, to the U.S. Treasury secretary, Lyman G. Cage, dated 22 August from the Norfolk Navy Yard in Virginia. Captain Munger referred to the commendation of Commodore J.C. Watson (USN) that credited the exemplary gunnery of the RC *Manning* at Cabanas, Cuba, and the multiple missions and duties of the cutter as an active naval vessel during the blockade of Cuba. The *Manning* guarded the supply and troop transport ships of the Fifth U.S. Army Corps at Santiago, Cuba, protected a U.S. Army supply base under Spanish attack, supported a shore landing party of sailors, and sent Dr. Mitchell, the *Manning* surgeon, to take care of wounded and ill U.S. Army personnel until army support arrived. On 14–15 July 1898 Captain Munger and designated officers visited insurgent camps and, with a flag of truce, a Spanish gunship, to inform them of the war-ending protocol and lifting of the naval blockade of Cuba.[41] The Treaty of Paris (December 10, 1898) formally ended the Spanish-American War and ceded Spanish colonial possessions to the United States. Commander Chapman C. Todd (USN) led a squadron of six U.S. Navy warships in the Cuban blockade. Cmdr. Todd complimented Capt. Munger (USRCS) in a letter to the U.S. Navy Department that stated, "I was associated with the *Manning* and Capt. Munger during the period of hostilities in the Northern Blockade ... and the high opinion I then formed of the efficiency of the *Manning* has been more than borne out by her service on the Southern Blockade which I had the honor to direct. I take great pleasure in calling attention to the highly meritorious services of this officer."[42]

Stephen Evans concluded, in his history of the USRCS in the Spanish-American War, that even though the postwar cutters had to "return to peaceful duties and cruising stations after the wartime missions of escort, blockade, scouting duty, landing operations and bombardments far from U.S. shores, the service had renewed its reputation as a blue-water, fighting outfit, played effectively on the Navy team, and won the respect of Navy commodores. This demonstration of military utility strengthened the position of the cutter branch in the federal system, and strengthened the nation's arms for the trials of force that lay ahead."[43]

The Spanish-American War influenced national and congressional support for the expansion for the U.S. Navy and U.S. Revenue Cutter Service. Thirteen revenue cutters served with the U.S. Navy, with the RC *McCulloch* battle in the Philippines at Manila Bay, and the RC *Hudson*, which towed a disabled U.S. Navy torpedo boat out of harm's way and enemy fire at Cardenas Bay in Cuba.[44] The USRCS proved itself worthy of its coordinated missions with the U.S. Navy, and its expanding role in national defense, as well as its significant domestic missions of marine safety, law enforcement, and national defense.

The United States Navy proved its worth in its national defense mission in the Caribbean and Pacific as it waged war in Cuban waters and in the insular waters of the Philippines with its steel ships and expanded global reach. The U.S. Navy established stations to protect American maritime commerce, in conjunction with the imperial expansion of America and in the future phases of Manifest Destiny. The writings of U.S. Navy scholar Alfred Thayer Mahan stimulated the support of Congress for the funding of expanded naval power. The combat successes of the Spanish-American War and the expanding U.S. Navy global presence was facilitated by U.S. Navy secretaries Benjamin F. Tracy (1869–1893) and Hilary A. Herbert (1893–1897) and a cooperative U.S. Congress, which provided the funding for five new battleships in 1895 and 1896.

U.S. Navy secretary John Davis Long and the assistant secretary and future U.S. president, Theodore Roosevelt, contributed significantly to the prewar naval preparations that served the nation so well in the Spanish-American War. The U.S. Marines illustrated their professionalism and combat readiness at Guantanamo Bay in Cuba. The naval warship capability was illustrated during the war when the battleship USS *Oregon* steamed from the West Coast of the United States, around Cape Horn at the southern tip of the Western Hemisphere, and into the victorious battle at Santiago Bay off the coast of Cuba.[45] The naval experience gained from the war and the expanded ship construction program prepared the U.S. Navy and U.S. Revenue Cutter Service for the joint U.S. naval combat missions in World War I in 1917 and 1918.

15

The USRCS and the U.S. Coast Guard Merge

With the end of the Spanish-American War the U.S. Revenue Cutter Service resumed its traditional domestic maritime missions in American interior and coastal waters. The global missions the Service performed alongside the U.S. Navy in that war prepared the Revenue Cutter Service for the evolving port security and national defense missions the USRCS and USCG would be assigned. War clouds were forming over the British Isles and continental Europe and would lead to the land and sea confrontations historians would call "The Great War," or World War I (1914–1918). In that time frame, the U.S. Revenue Cutter Service would merge with the U.S. Life-Saving Service to form the United States Coast Guard in 1915. That date should not be confused with the origin of the U.S. Revenue Marine in 1790, the predecessor agency of the U.S. Revenue Cutter Service and then the U.S. Coast Guard of 1915.

Captain-Commandant Ellsworth Bertholf would lead the Revenue Cutter Service and then the U.S. Coast Guard into World War I in joint missions with the U.S. Navy. In the post–World War I period Commodore Bertholf fought to maintain the autonomy of the United States Coast Guard and resisted the political forces that tried to abolish the USCG by combining that naval service with the United States Navy. Bertholf's background indicated the courage and leadership qualities that would propel him into national prominence before he was appointed commandant of the United States Revenue Cutter Service (1911) and the United States Coast Guard. The United States entered World War I belatedly in 1917, with the massive infrastructure and naval and military power that contributed to the end of the war, the 11 November 1918 armistice, and the controversial Treaty of Versailles signed in France on 28 June 1919.

Bertholf commenced his service with the USRCS as a cadet at the Revenue Cutter School of Instruction in 1885. Upon graduation Ensign Bertholf

169

was assigned to the RC *Woodbury* (1887) and commissioned a third lieutenant in 1889 with the completion of two years of sea service.[1] The adventuresome officer served off the coasts of the United States, and most extensively in the challenging seas and weather of territorial Alaska and the Bering Sea. He was a Congressional Gold Medal recipient for his leadership in the dangerous 1897–1898 Alaska Overland Expedition rescue of stranded American whalers off Point Barrow. And if that polar geography was not enough of a challenge, Bertholf travelled on a treacherous mission by dog sled across the frozen landscape of Russian Siberia in 1901 to acquire reindeer herds for starving Inuit natives in northern Alaska. Captain Bertholf then assumed command of the USRC *Bear* on the Bering Sea Patrol.

In 1911 Captain-Commandant Bertholf was appointed the head of the U.S. Revenue Cutter Service in the administration of President Woodrow Wilson and was reappointed commandant of the fledgling United States Coast Guard when the U.S. Revenue Service was combined with the U.S. Life-Saving Service in 1915. The successful merger was solidified by Bertholf's leadership skills and knowledge. In 1912 Captain Bertholf served as an American representative to the International Conference on Safety at Sea in London, England. That meeting of key maritime nations led to the creation of the International Ice Patrol and the tracking and charting of icebergs with radio warnings of their presence in the high latitude Atlantic sea lanes to protect the safety of ships, crews, and passengers. In addition to his responsibilities as Commandant of the U.S. Coast Guard, Captain Bertholf served on the Board on Vessel Movements and Anchorage and on the North Atlantic Board of International Ice Observation and Patrol.

Commandant Bertholf was appointed to the rank of Commodore in World War I, the first USRCS and USCG officer to achieve what the naval services call Flag Rank. Bertholf achieved another first in his career in the Revenue Cutter Service when he was admitted to the U.S. Naval War College at Newport, Rhode Island. On 30 June 1919 Commodore Bertholf retired from the United States Coast Guard and continued his contributions to maritime history and seafaring, with his work as a vice president of the American Bureau of Shipping.[2] In 1921 Ellsworth Price Bertholf passed away in York City at the age of 55. The commodore ;was buried at Arlington National Cemetery with full military honors.[3] The Legend Class National Security *Bertholf* (WMSL-750) honors the memory and contributions of the first Coast Guard commandant and commodore. The CGC *Bertholf* was launched in 2006, operates in the Pacific Ocean, and is based in Alameda, California. The cutter missions include national defense and all of the other multi-mission responsibilities of the United States Coast Guard.

A complex mixture of international causes that included the clash of alliance systems, military and naval armament competition, and nationalistic and colonial/ territorial quarrels between the European nations caused World War I. Those conflicts caused the nations to restrict trade between the United States and the belligerent powers. Pro-British bias on the part of America caused the Wilson administration to favor the United Kingdom and to engage in trade with Britain. The German monarchy rejected America's claims of neutrality and sovereign rights on the high seas, and German submarines (U-boats) sank Western ships. Unrestricted German submarine warfare threatened U.S. sovereignty, and forced Congress and the president to declare war on the Central Powers of Germany, Turkey, and the Austro-Hungarian Empire. U.S. naval and military forces were sent overseas late in the war (1917) to join the Allied nations in their fight against Kaiser Wilhelm, the German ruler, and the other

Captain-commandant of the U.S. Revenue Cutter Service and then the U.S. Coast Guard (1915), Ellsworth P. Bertholf led the U.S. Revenue Cutter Service merger into the new U.S. Coast Guard and then guided the U.S Coast Guard into World War I at home and overseas.

Central Powers. The U.S. Navy and U.S. Coast Guard escorted merchant ship convoys across the Atlantic Ocean and into the Mediterranean Sea. During the war the Coast Guard was responsible for policing the heavy shipping traffic that passed through American seaports and to Europe. Cutters served as convoy escorts and conducted antisubmarine warfare patrols. The USCGC *Tampa* was sunk by a German submarine[4] with heavy loss of life. The convoy and ASW experience gained by the USCG in World War I would prepare the Service to support the U.S. Navy in port security and naval combat operations in World War II (1939–1945).

The significant role played by the USCG in search and rescue (SAR) and national defense missions influenced Congress to fund the establishment of the first U.S. Coast Guard Aviation Stations at Clouchester, Massachusetts, and Cape May, New Jersey, and subsequently other stations on the ocean coasts and Great Lakes. The Coast Guard aviation force assisted the cutters in SAR, then Prohibition enforcement (1920–1933),[5] and SAR, ASW, and reconnaissance missions during World War II. The European naval and military armament race influenced American strategic planning. The U.S. would sponsor post–World War I disarmament conferences that attempted to limit the building of naval vessels. The future World War II antagonists would ignore the respective treaty provisions. Euro-American attempts to abide by the provisions would put them at a power disadvantage when Germany and Japan embarked on the aggressive paths that led to World War II. Before World War I the United States participated in the buildup of its military and naval strength to compete with Europe in a balance of power contest among the maritime nations. The U.S. was prepared to protect the imperial colonial territories it gained in the Caribbean and Pacific from the Spanish-American War of 1898. The imperial nations of Europe and the increasingly powerful nation of Japan were preparing to do the same: acquire and protect colonial possessions.

Japan's acquisition of insular and continental territories led the leaders of that nation, and the United States and Europe, to compete with each other and prepare for the possibility of conflicts. In that regard, the United States developed Plan Orange in 1911, a secret contingency operation for possible war with Japan. The U.S. launched an expensive program to build gigantic warships called dreadnoughts. Britain built its first huge warship, appropriately named the HMS *Dreadnought*, in 1906. America followed in 1910 with the USS *Michigan* and planned for 13 more such combat vessels over the next several years. The USS *Idaho* (1908) carried four 12-inch guns. The USS *Pennsylvania* (1916) carried twelve 14-inch guns that could fire projectiles out to a range of 12 miles. The significance of the shift from coal to oil as a power source influenced naval shipbuilding, strategy and tactics. The USS *Michigan* (1910) burned only coal, but two of the other dreadnaughts commissioned that year could burn coal or oil. The battleship USS *Nevada* (1916) was strictly an oil burner and could cruise faster and farther than coal-powered warships, with a speed increase of up to 20 percent.

The buildup and modernization of the U.S. naval fleet was fortuitous. On 12 March 1917, German U-boats torpedoed the U.S. merchant steamship *Algonquin*. That same month German submarines sank several other U.S. ships. This assault on U.S. sovereignty and territory caused President Woodrow Wilson, who had campaigned on the slogan, "He Kept Us Out of War," to ask the Congress for a war declaration on 2 April 1917. On 6 April 1917, Congress provided that declaration to the president for his signature.[6] The U.S. Navy and U.S. Coast Guard joined forces to guard merchant ships bringing wartime supplies to Great Britain. The operation involved a convoy system of merchant ships protected by armed naval vessels. The USN destroyers and battleships did not participate directly in combat alongside the British Royal Navy against the German fleet. For the most part, the German fleet was confined to the northern waters and to ports of continental Europe. The USN guarded convoys and ports and laid mines under the command of Captain Reginald R. Belknap (USN), who provided a fleet of mine laying cruisers and former merchant steamships for patrols in British waters in 1918. The mines eventually totaled 56,600 in number. The floating devices were shipped from Norfolk, Virginia, and anchored in the North Sea. The mine laying was extensive and complex. It was not cost effective, but it did keep German U-boats out of strategic waters and sank at least one enemy submarine.

The Royal Navy of the United Kingdom and the British Admiralty requested that the USN bring only coal burners to Europe, given the difficulty the UK had acquiring petroleum. The British ships were mostly coal burners. In the end, the USN sent six of its oldest coal-burning battleships to European waters under Rear Adm. Hugh Rodman—USS *Arkansas, Delaware, Florida, New York, Texas* and *Wyoming*—to Scapa Flow off the coast of northeast of Scotland in December 1917. The U.S. naval force was called Battle Squadron Six of the Grand Fleet. The USS *Utah, Oklahoma* and *Nevada* were stationed off the west coast of Ireland.

The naval element that did the most significant and influential fighting in Europe was the United States Marines under Major Gen. George Barnett, and the combat commands of Major Gen. John A. Lejeune (USMC) and Brig. Gen. Charles A. Doyen. The Marines cooperated with Gen. John J. Pershing (U.S. Army), the head of the American Expeditionary Force (AEF). U.S. Army and U.S. Marine Corps troops suffered heavy casualties in bloody fighting against German troops, and were instrumental in securing the eventual Allied victory and the Armistice of 11 November 1918.[7]

The diversity and complexity of naval war technology, ships, tactics and strategy in the World War I period is illustrated by the range of boats and ships used by all sides in the maritime conflicts. The hulls of battleships like the USS *New Jersey* were painted with camouflage patterns to make the ships blend in with the water and sky to ostensibly be less visible to enemy vessels. But up close, the patterns were so striking that one could wonder if in fact the artistry worked. The German navy was advanced in the use of "Untersee Boots" (Under Sea Boats, or U–Boats) that were propelled by fuel oil and battery power. American battleships at Scapa Flow did not see real combat action, but destroyers and sub-chaser patrol boats protected harbor areas and participated in naval activity. Harbor regions and bases included busy ship repair and resupply facilities.

In April 1917 the USN possessed 361 vessels, 151 of which were warships. The Allies needed fast, maneuverable warships and merchant vessels to carry ordnance, general supplies and troops. The destroyers protected merchant ship convoys from U-boat attacks. The U-boats would surface, use deck guns to sink unarmed merchant vessels, and submerge to use torpedoes on armed merchant ships and warships. In 1918 three hundred thousand U.S. troops arrived at French ports every month. The convoy system was successful to the extent that just six troop transports were sunk by U-boats during the war. Four of those vessels were carrying no troops or cargo.[8] U.S. oil tankers fueled warships at sea. After the war the U.S. Navy Transport Department calculated that naval transports carried one million troops to European waters and shores. In 1917 a U-boat sank the USS *Alcedo*. By the middle of 1918 a half-dozen U-boats had crossed the Atlantic to operate in U.S. waters. In 1917 U.S. aircraft landed in France, and four U.S. submarines docked in the European Theater of Operations in the Azores Islands. In November of 1917 a U.S. warship sank a U-boat. The American convoy system was an overall success. By 1918 the Allied convoy system had reduced merchant ship losses to German submarines in the Atlantic maritime realm.[9]

In the Second World War, this "Battle of the Atlantic" would resume in a contest between the Allies: the U.S. Navy and Coast Guard and Royal British and Canadian navies against an even more extensive and successful German U-boat campaign that would eventually be defeated by Anglo-American air and sea attacks. Merchant convoys were protected by naval escorts, but high casualty rates were suffered by both Allied and German seamen.

Coast Guard historian Irving H. King wrote that the decision by the U.S. Congress to maintain the USCG as "an independent seagoing service in the Treasury Department [rather than combining the service with the USN after the Spanish American War] was vindicated by [the Coast Guard's] contribution to the Allied success during World War I in the Atlantic Ocean, Mediterranean Sea, and at major U.S. ports. Coast Guardsmen reduced and prevented losses from sabotage and careless explosive handling."[10] Coast Guard officers commanded U.S. Navy warships and saved the lives of mariners and ship passengers during the U–Boat onslaught against merchant and passenger vessels. That combat experience and contribution to national defense paved the way for the USCG to subsequently partner with the U.S. Navy in future American wars overseas, including World War II, Korea, Vietnam, and the Persian Gulf.[11]

Rear Admiral Frederick C. Billard would serve as the commandant of the Coast Guard from 1924 to 1932, during the challenges of the federal enforcement of Prohibition laws (1920–1933). Prior to that, Captain Billard served as the superintendent of the U.S. Coast Guard Academy, and commander of the USCG training cutter *Onondaga* out of New London, Connecticut. The USCGC *Onondaga* transported newly commissioned Coast Guard ensigns to the ocean-going cruising cutters being prepared for service in European waters in World

War I. The *Onondaga* escorted other vessels to Coast Guard units, including the Great Lakes cutter *Mackinac* to the Atlantic Coast and Long Island Sound off of New York City.[12] Captain Billard received orders to report to London, England, to command the 302-foot USS *Aphrodite*, which had a crew complement of 130 officers and men and the mission assignment of escorting merchant vessels off the coast of France. After the 11 November 1918 armistice that ended World War I combat, the *Aphrodite* struck a German mine in the North Sea, but the crew completed the mission and Captain Billard returned to the United States.[13]

Harry G. Hamlet would serve as Coast Guard commandant from 1932 to 1936. Prior to that appointment, Hamlet commenced his career as a cadet at the Revenue Cutter School of Instruction in 1894, trained on the USRC *Chase*, earned his ensign commission in 1896, and was assigned to the USRC *Bear*. Hamlet was invited to attend the U.S. Naval War College in Newport, Rhode Island. He was assigned to organize personnel training at two naval bases after the United States entered World War I, was then attached to U.S. Naval Forces in France, and commanded the U.S. Navy warship *Marietta* in 1918. Captain Hamlet and his crew rescued the crew of the USS *James* as that ship was sinking in heavy seas off the coast of France in 1919. Captain Hamlet and his crew

The 206-foot USRC *Gresham* served as a convoy escort across the Atlantic in World War I after the United States entered the war in 1917.

Top: The 205-foot USRC steamer *Manning* served as a blockade and escort vessel in the Spanish-American War. During World War I the cutter served under the U.S. Navy as an Atlantic convoy escort and then out of Gibraltar in the Mediterranean Sea. *Bottom:* Another view of the USRC *Manning* steaming out to sea.

The 206-foot U.S. Revenue Cutter *Onondaga* served as the Revenue and Coast Guard Academy training ship for cadets after being designated the United States Coast Guard cutter *Onondaga* in 1915. The cutter enforced wartime neutrality laws in the Atlantic and Chesapeake Bay maritime region and transported Coast Guard personnel and civilian officials to various sites and assignments on the Atlantic coast during World War I.

exhibited exemplary seamanship in high waters under gale-force winds. For that rescue, Hamlet received the Congressional Gold Life-Saving Medal from U.S. Treasury Secretary Carter Glass and the Silver Star and Special Commendation from U.S. Navy secretary Josephus Daniels.[14]

Admiral Russell R. Waesche served as commandant of the United States Coast Guard before and during World War II, from 1936 to 1946. Admiral Waesche prepared the USCG for the war and articulated training and missions with the U.S. Navy, U.S. Marines, U.S. Army, and the U.S. Army Air Force. For his leadership skills and contributions Adm. Waesche received commendations and awards from President Franklin Roosevelt, the top civilian and military leaders of the federal government, leaders of the other U.S. Armed Forces, and leaders of foreign nations. Waesche's memory and achievements were commemorated when the 418-foot Legend Class National Security Cutter *Waesche* (WMSL-751) was commissioned in 2010.[15] Russell R. Waesche exhibited his leadership skills early in his career. From his graduation as an ensign from the

U.S. Revenue Cutter School of Instruction in 1906 to achieving flag ranks in the 1930s, Waesche exhibited superior leadership and administrative skills.

President Woodrow Wilson anticipated the high probability of the chance of war between the United States and Germany after 1914 and was aware of that nation's submarine (U-boat) threat. Wilson created the Interdepartmental Board on Coastal Communications in 1916, one year before the United States entered the European war against the Central Powers. The board was ordered to assess the Atlantic coastal defense systems, military stations, communications networks, search and rescue capabilities, and the saving of life and property on land and sea. The Coast Guard was tasked with upgrading its telephone communications systems and networking with shore stations, lighthouses, lightships, and lifesaving (lifeboat) stations. The placement and testing of the necessary lines, circuits, copper wiring, and technical transmission facilities were extraordinarily complex tasks.

Captain Alex R. Larzelere, USCG (Ret.), the author of an exemplary history of the United States Coast Guard in World War I, described how the U.S. Navy ordered the entire coastal communications system placed under the authority and supervision of the Coast Guard commandant. First Lt. Russell R. Waesche was promoted to captain and placed in charge of the communications infrastructure. He increased the complement and training of civilian, enlisted, warrant and commissioned officers to operate and maintain the communications systems and make them an essential and effective element of national defense.[16]

Admiral James Francis Farley served as commandant of the U.S. Coast Guard from 1946 to 1950. In 1912 Cadet Farley graduated from the Revenue Cutter Service School of Instruction, the predecessor of the U.S. Coast Guard Academy. Third Lt. James Farley advanced in rank from 1912 to 1915, when he saw service on the Coast Guard cutters *Mohawk, Seminole, Onondaga*, and then the *Yamacraw* when that cutter was assigned to the U.S. Fleet Patrol. During World War I Lt. Farley sailed on convoy escort patrol in the Mediterranean Sea between the British colony of Gibraltar and various ports in the United Kingdom. For his distinguished service in the Great War, Farley earned the Victory Medal and Escort Clasp.[17]

As have Coast Guard commandants before and after him, Admiral Robert J. Papp, Jr. (2010–2014), represented the U.S. Coast Guard at numerous events, facilities, and missions. On Veterans Day, 11 November 2010, Homeland Security Secretary Janet Napolitano and Adm. Papp commemorated the Coast Guard mission in World War I. The occasion was a wreath-laying ceremony at the U.S. Coast Guard Memorial in Arlington National Cemetery in Virginia. The Coast Guard Memorial on Coast Guard Hill was built to honor the combat crew members lost when the USCGC *Seneca* and USCGC *Tampa* were torpedoed in 1918. The World War I Memorial lists the names of all of the U.S. Coast Guard personnel who perished in the Great War.[18] During the war Coast Guard cutters and crews performed SAR and antisubmarine warfare (ASW) missions, dropping depth charges on suspected German U-boat locations. On search and rescue missions civilian crews were rescued from damaged or sunken vessels in flaming petroleum-soaked waters. Within U.S. territorial and coastal waters the USCG managed port security under Coast Guard officers designated as captains of the port (COTP). Specially trained crews guarded against enemy espionage and sabotage, supervised dangerous explosive loading on and off military and civilian transport ships and railroad trains, and cooperated with civilian firefighting teams in port and on shipboard fires. Members of the U.S. Lighthouse Service (USLHS) teamed up with the USCG to administer aids to navigation (ATON) units, including lightships, lighthouses, and buoy systems.[19]

Captain Alex Larzelere, USCG (Ret.), described the interrelationship of the USCG and the U.S. Navy in the Great War. Larzelere chronicled how the Navy acquired ships from a variety of government agencies and private owners. The USN requested that the USCG assign experienced enlisted personnel and officers to operate and command a variety of purchased and donated yachts and ships. In 1917 the U.S. Navy Bureau of Navigation asked USCG Headquarters in Washington, D.C., to place Coast Guard officers in command of USN ships and other large vessels and yachts.[20] Coast Guard officers were assigned to some of the largest warships in the U.S. Navy and saw combat action on those vessels. Three Coast Guard offices were aboard the 413-foot *Minneapolis*. First Lt. Leon C. Covell (USCG) was a navigator in the 477-crew complement. The cruiser carried eight 4-inch guns, two 6-inch guns, an 8-inch gun, and four 18-inch guns. The ship escorted vessels across the Atlantic and patrolled off the eastern seaboard of the United States.

The USS *San Diego* led Atlantic escort patrols to Europe. One and a half months after 2nd Lt. Henry G. Hemmingway (USCG) came aboard the vessel in 1918, the *San Diego* was steaming between Portsmouth, New Hampshire, and New York Harbor and struck a German mine. The huge 504-foot warship, carrying 38 guns ranging from 2 to 8 inches, sank within 30 minutes. Of the 1,114 crewmembers, only six were lost at sea. The USS *San Diego* was, according to Larzelere, the largest U.S. warship lost in World War I. Lt. Hemmingway survived the incident and was reassigned to another naval vessel. Navy minesweepers found six enemy mines in the vicinity where the USS *San Diego* sank,[21] which illustrated the extent, range, and activities of German U-boats off the U.S. Atlantic Coast. The contribution of the U.S. Coast Guard to domestic security and national defense during the war was described by U.S. Navy secretary

Josephus Daniels and cited by Larzelere: "The professional ability of the Coast Guard officers is evidenced by the fact that 24 commanded combat ships in European waters, five vessels in the Caribbean, and 23 attached to naval districts. The Navy Department assigned to command those ships only officers whose experience and ability warranted the [assignment], and in whom the Department had implicit confidence."[22]

The commander of the U.S. Coast Guard's New York Division, and Captain of the Port (COTP), Captain Godfrey L. Carden (USCG), exemplified the professional leadership the Coast Guard provided at home and overseas. Captain Carden operated with the U.S. Navy but reported to U.S. Treasury secretary William Gibbs McAdoo. Captain Carden's strict, and often unpopular, regulations governing the loading, unloading, and movement of ammunition ships in the harbor of New York City[23] no doubt spared the region untold costs in explosion, fire disasters and human casualties. The significance of the improved port administration was indicated by the fact that ship and warehouse fires and explosions did occur in the harbor and port regions of New York and New Jersey, but were successfully responded to, and contained by civilian and Coast Guard firefighters, many of whom earned commendations for their heroic responses.[24]

Larzelere did not neglect the significance of the Great Lakes region in the flow of ships and supplies for the war effort and the significance of the Coast Guard in the region of the Inland Seas in their role in protecting commerce and providing for the national defense. The Espionage Act of 1917 gave the Coast Guard the responsibility for the enforcement of laws and regulations pertaining to the movement and anchorage of ships in strategic navigable waters, transferring that enforcement mission from the U.S. Army Corps of Engineers to the U.S. Department of the Treasury and the U.S. Coast Guard. Treasury Secretary William G. McAdoo assigned Coast Guard officers to security responsibilities at the strategic ports of New York City; Hampton Roads, Norfolk, Virginia; Philadelphia, Pennsylvania; and Sault Ste. Marie, Michigan, at the Soo Locks on the Great Lakes.[25] Eventually, a COTP would be assigned to the Twin Ports of Duluth, Minnesota, and Superior, Wisconsin, at the southwest terminus of Lake Superior. The commander of U.S. Forces in France was pleased with the assignment of Capt. Detlef F. A. de Otte as the port security commander at the port of Brest. The infrastructure and transportation complexities at Brest were enormous, and significant in the European war effort. The duties of Capt. de Otte included facilitating operations for vessel oiling, watering, and coaling; transporting troops; handling a variety of watercraft including ships, barges, boats, and tugs; assigning dock and port anchorage sites; and cooperating with the maritime and military authorities of France. Between

1917 and 1919, administrative services had been provided for more than 350 supply, combat, and transport ships, at an estimated tonnage in excess of 6,800,000 at the port of Brest. Supply ships delivered more than 257,000 tons of cargo. The transport ships moved more than 700,000 troops.[26]

Larzelere did a thorough analysis of the significance of the Great Lakes in the Great War, as "the vital waterway connecting the internal commerce of the Great Lakes to the Atlantic Ocean that was a major concern"[27] to navy secretary Josephus Daniels in his annual report. Daniels concluded, "The tremendous amount of shipping that passes from upper to the lower lakes through the canals, locks, and waterways of the St. Mary's River system makes it most important to carefully regulate this traffic,"[28] given the potential of the locks for sabotage and infrastructure damage from collisions and winter ice. The USCGC *Mackinac* (Capt. Edward S. Addison, USCG) patrolled the strategic Soo Locks area, assisted by Capt. de Otte in command of the USCGC *Morrill* out of Detroit, Michigan, before World War I erupted. Captain Otte was then transferred to Europe. Captain Addison was appointed COTP at the Sault Ste. Marie port. The U.S. Army Corps of Engineers and armed U.S. soldiers guarded the Soo Locks during the war.[29] The *Mackinac* was ordered on 10 November 1917 out of Sault Ste. Marie to the port of New York City to guard the anchorage region of ships and explosive loading activities.[30]

After the war ended, President Woodrow Wilson attended the Paris Peace Conference (1919), where he contributed to the concepts enunciated in the Versailles Treaty of 28 June 1919, including the formation of the League of Nations. The president and other political leaders returned from Europe on the USS *George Washington*. From that U.S. Navy ship the president and his official entourage transferred from the large navy transport onto the smaller cruiser USCGC *Ossipee*, under the command of Captain William H. Munter, who had commanded the cutter in the wartime European maritime theater. Captain Munter and his crew transported President Wilson and his distinguished party to Commonwealth Pier in Boston, Massachusetts. The notables who accompanied President Wilson on the USCGC *Ossipee* included his wife, Edith Wilson, Assistant (Navy) Secretary Franklin D. Roosevelt, and his wife, Eleanor Roosevelt. Back in Washington, D.C., Commodore Ellsworth P. Bertholf, the Coast Guard commandant, appeared before a congressional hearing on 6 February 1919 to speak against a bill that would have kept the U.S. Coast Guard permanently under the jurisdiction of the U.S. Navy.

When President Wilson returned to Washington, he spoke with the leaders of Congress to explain the Treaty of Paris, argue for its ratification by the U.S. Senate, and explain to a dubious Congress why he wanted the United States to join the new League of Nations. The president also had to decide whether to approve the joint resolution of Congress to return the United States Coast Guard to the Department of the Treasury. U.S. Navy secretary Josephus Daniels and the assistant secretary of the U.S. Navy, Franklin Roosevelt, opposed the resolution and favored maintaining the USCG under the USN. Commodore Bertholf had previously argued against the absorption of the USCG by the USN, correctly concluding that the Coast Guard would then no longer exist. The recently appointed Treasury secretary, Carter B. Glass, and Commodore Bertholf strongly opposed the proposed absorption of the U.S. Coast Guard into the U.S. Navy.[31] Many Coast Guard officers favored the transfer of the Service to the U.S. Navy, considering the elements of rank, pay, privilege and prestige the U.S. Navy and congressional sponsors enunciated. Coast Guard enlisted personnel would not have fared as well, given the rather unequal formulations offered to rated men, lifesavers, and surf sailors.[32] Captain Richard H. Laning, Chief of the U.S. Navy Bureau of Operations, testified that the expanded U.S. Navy needed the experienced Coast Guard officers who had performed so well for the navy during the war. Captain Laning said the high quality of USCG officers would be difficult to replace after the war, and that quality could otherwise come only from cadets and future officers matriculating out of the U.S. Naval Academy at Annapolis, Maryland. The U.S. Navy needed more quality officers immediately

Further congressional hearings, extensive proposals, resolutions, and the testimony of political and naval opponents and proponents of the proposed absorption of the USCG by the USN continued unabated until Secretary Carter Glass met with President Wilson just before the president left for his national tour to stir up support for U.S. membership in the fledgling League of Nations. Secretary Glass persuaded Wilson to act on the joint congressional resolution before commencing his trip. President Woodrow Wilson then issued Executive Order Number 3160, on 28 August 1919. The United States Coast Guard was thereby returned to the jurisdiction of the U.S. Department of the Treasury,[33] where the Service had been since its birth in 1790 as the U.S. Revenue Marine.

16

An Overview of Naval Operations from 1790 to the 20th Century

The evolution of the maritime missions of the U.S. Revenue Marine, Revenue Cutter Service, and U.S. Coast Guard, often in joint missions with the U.S. Navy, have been considered in previous chapters. A final overview will be considered in this chapter to synthesize, clarify, and add to those missions of the Coast Guard and its institutional agency predecessors at home and overseas. The missions involve economic and resource protection, saving life and property, search and rescue, port security, aids to navigation, ship inspection, scientific exploration, and national defense.

The immediate predecessor agency of the U.S. Coast Guard was the U.S. Revenue Marine, later called the U.S. Revenue Cutter Service. The Revenue Marine was founded on 4 August 1790, during the presidency of George Washington. Congress created this initial United States navy. The institutional United States Navy and Federal Navy Department were formed when Congress provided the funding for the U.S. Navy in 1795 and put the first USN vessels in the water in 1798. The United States Navy and United States Marine Corps origins date back to colonial America and the Revolutionary War. The Revenue Marine was placed by Congress under the jurisdiction of the United States Department of the Treasury and Secretary Alexander Hamilton in 1790. The small sea service was simply referred to as "The Cutters," or "A System of Cutters."

The early cutter officers came from the merchant fleets and the disbanded Continental Navy after the Revolutionary War. The first cutter commander was Captain Hopley Yeaton, who hailed from the maritime state of New Hampshire. The first revenue vessels were small boats carrying one or two masts and included the Revenue Cutters *Massachusetts*, *Diligence*, *Virginia*, and *Pickering*.

The cutters would gradually increase in size and armament into sloops, schooners, and brigs, and the numerous boats and craft the Service used in its myriad of missions. Revenue cutters, and future Coast Guard cutters, were commonly named after U.S. Treasury secretaries, and later after Coast Guard luminaries. During the Lyndon Johnson presidency of the 1960s, the USCG would be placed under the Department of Transportation (DOT). In 2003, the Coast Guard was transferred to the new Department of Homeland Security (DHS).

The cutter fleet distinguished itself in partnership with the U.S. Navy in the Quasi-War with France (1797–1801), the "Second War of Independence" against Britain in the War of 1812, and wars against pirates, slave ships, and the Seminole Indians of Florida. In the 19th century the Revenue Marine and Revenue Cutter Service would team up with the U.S. Navy in the Mexican-American War (1846–1848), Civil War (1861–1865), and Spanish-American War (1898) and again with the U.S. Navy in domestic and overseas missions as the United States Coast Guard in World War I (1917–1918).[1] The coordination of U.S. Revenue Cutter Service and U.S. Coast Guard missions with the U.S. Navy has been part of United States naval history from 1790 to the present. The periodic union of the two naval services dates from the act of the U.S. Congress on 2 February 1799 that stated, "The President of the United States shall be, and is hereby authorized to place on the naval establishment, and employ accordingly, all or any of the vessels, which, as revenue cutters, have been increased in force and employed in the defense of the seacoasts ... and thereupon, the officers and crews of such vessels, may be allowed, at the discretion of the President of the United States, the pay, subsistence, advantages and compensations, proportionately to the rates of such vessels, and shall be governed by the rules and discipline which are, or which shall be, established for the Navy of the United States." In addition, the act of March 2, 1799, provided that the cutters "shall, whenever the President of the United States shall so direct, cooperate with the Navy of the United States, during which time they shall be under the direction of the Secretary of the Navy."[2]

Eight revenue cutters patrolled the Gulf Coast and West Indies regions in the Quasi-War with France between 1798 and 1801. The cutters carried from 10 to 14 deck guns and crew complements ranging from 30 to 70 officers and men. United States naval forces captured more than 20 enemy vessels. The RC *Pickering* captured 10 enemy vessels in West Indies and Caribbean waters. When the U.S. Navy was sent to the Mediterranean Sea to battle the Muslim Barbary pirates between 1800 and 1815, revenue cutters did not sail to African waters, but several officers of the Revenue Marine joined U.S. Navy forces to battle Barbary pirates that captured American merchant ships and imprisoned and enslaved crew members.

The War of 1812 against Britain was fought in U.S. coastal waters, on the high seas, in U.S. territory, and in Canada. The smaller cutters did coastal, river and shallow- and littoral-water patrols. The larger, deep-draft U.S. Navy warships were better suited for deep- (blue) water patrols and combat. Cutters engaged British merchant ships and naval frigates and captured several while, on a few occasions, being outgunned. Some cutters were captured by enemy forces. Among the Revenue Marine standouts were the valiant crews of RC *Vigilant,* RC *Surveyor,* and RC *Eagle* the latter off Long Island, New York. Between the War of 1812 and the year 1820 the Revenue Cutters *Louisiana* and *Alabama* battled pirates on land and sea in the Caribbean and Gulf maritime areas and the Louisiana bayous and swamps. The pirates were eventually subdued or driven to Latin American sites, but they continued to ravage U.S. territory although in gradually diminishing attacks.

The revenue cutters supported amphibious operations with the U.S. Army and U.S. Navy in the 1830s and 1840s in the coastal and swamp regions of Florida. Cutters carried military and naval personnel, ordnance (small arms, artillery and munitions), and other supplies, plus government dispatches. The cutters and armed crews blocked river passages and defended settler settlements. In the United States–Mexico War (1846–1848), the Revenue Cutter Service supplemented U.S. Navy operations off Cuba and strategic ports in amphibious, rescue, supply, and combat missions. In the 1850s, 2nd Lt. James E. Harrison, on the RC *Jefferson Davis,* engaged in combat operations with U.S. Army Infantry units against Native Americans in the territory of Washington in the Pacific Northwest. Lt. Harrison led ground forces in battle.[3]

The period between the origins of the U.S. Revenue Marine and World War I exhibited a significant evolution from sailing (wind-powered) vessels to auxiliary- (sail and coal/steam) and petroleum-powered naval ships, and from wooden-hulled U.S. Navy ships and U.S. Revenue cutters to iron-clads (iron over the wooden hull), and then steel-hulled ships. Enlisted men and commissioned and warrant officers had to learn increasingly complex elements of seamanship and mechanical engineering and apply those elements to maritime operations. Domestic politics, international diplomacy, and geopolitical and colonial rivalries between seafaring nations and their military and naval missions changed along with infrastructures, foreign relations and international exigencies. The U.S. Revenue Marine, Revenue Cutter Service, Coast Guard and Navy mirrored these changes, and were called upon to defend and protect the United States and its interests at home and overseas.

The Civil War (1861–1865), in both the Union and the Confederate States of America, saw rapid technological change in political, military, and naval institutions. To match the overwhelming population, industrial, and infrastructure advantage of the Union states, the Confederacy had to respond with overseas trade and purchases of supplies and war technology, capture Union ships, convert commercial vessels into warships, and station Confederate political and military officials in Britain to engage in diplomatic subterfuge and the purchase of vessels from British shipyards to wage commercial and naval warfare against the Union. Southern military and naval leaders perfected the use of iron monitors and iron-clad vessels with submerged bow rams to attack and sink Union blockading vessels. Confederate commerce raiders escaped the Union naval blockade to conduct trade and force the strategic, global repositioning of Union warships to confront, capture and destroy Rebel naval and civilian raiders. The maritime conflicts occurred along rivers, lakes, and ocean coasts; the Gulf of Mexico in North America; and across the Atlantic Ocean into European, Asiatic and even polar waters.

The Confederate States Navy (CSN) and its Confederate States ships were designated CSS vessels, as opposed to the U.S. Navy and its USRCS and USN vessels, designated as USRCS and USS vessels. The CSN developed maritime explosives (mines or "torpedoes") and designed and used the first submarine, the CSS *Hunley,* in naval combat. Donald L. Canney did extensive research and writing on the Confederate steam navy and its warships, technology, and commerce raiders. Canney described and illustrated 19th-century naval technology, foreign-built warships, fortifications, floating batteries, explosive maritime mines (torpedoes), and submersible vessels. Canney described the administrative skills of the Confederate States Navy secretary, Stephen Mallory, and the scientific and oceanographic genius of Commander Matthew F. Maury (USN/CSN) and his innovative naval inventions and acquisitions.[4]

During the Civil War the CSA claimed sovereign status and therefore referred to the conflict as the War Between the States, or the War of Northern Aggression. Northerners referred to the conflict with the South as the War of Secession and, because of the struggle within the nation between its citizens and even within families, referred to the conflict as the Civil War. The conflict pitted the Federal navy and military of the North against the Confederate forces of the South that had seceded from the Union. The Confederate secessionist movement was activated by the 1860 election victory of President Abraham Lincoln. The CSA was headed by a former West Point graduate, Mexican-American War veteran, and U.S. secretary of war, Jefferson Davis. During the initial secessionist period, several U.S. revenue cutter commanders turned their vessels over to the Confederacy.

The USRC/USS *Harriet Lane* fired the first naval shots of the Civil War in the harbor at Charleston, South Carolina. The Confederate firing on the Union Fort Sumter on 12 April 1861 was the immediate incident that started the Civil War. The *Harriet Lane* participated with the U.S. Navy at Hatteras Inlet and associated fortifications on coastal North Carolina. After the tragic assassination of President Lincoln on 21 April 1865, U.S. revenue cutters were ordered to search outgoing vessels to prevent the escape of Lincoln's killers.[5] The *Harriet Lane* served with the U.S. Navy in the battle up the Mississippi River from the Gulf, past formidable Confederate fortifications and joined U.S. Marine, U.S. Army, and U.S. Navy forces upriver to capture New Orleans, Louisiana, and also served in the U.S. Navy blockade along Confederate seacoasts and rivers.

The Confederate government attempted to break the Union naval blockade with fast blockade-runners that transported cotton and tobacco to sell overseas and brought supplies and ships from Europe and European colonies. Among the purchased blockade-runners was the side-wheel steamer *Lady Sterling*, originally a British vessel. The USS *Eolus* captured the *Lady Sterling* on its first voyage out of Wilmington, North Carolina, after a brief round of gunfire that damaged the privateer and forced it aground. The U.S. Navy then commissioned the captured privateer and used it as a blockade vessel and, after the war, as a yacht for federal officials.[6]

In the Spanish-American War (1898), eight revenue cutters carrying 43 guns supported Rear Admiral William T. Sampson (USN) in the blockade of Havana, Cuba. The USRC *Manning* fired on Spanish fortifications off the coast of Cuba in May 1898. The USRC *McCulloch*, with a crew complement of 105 and six deck guns, was at the Battle of Manila Bay in the Philippines and was attached to Admiral George Dewey as his communications and message dispatch ship. Off the Cuban coast in May of 1898 the RC *Hudson* (Lt. Frank H. Newcomb) engaged in heavy combat with Spanish shore batteries and gunboats to rescue the damaged USS *Winslow* and the surviving crew. For this gallant combat action and heroism, the U.S. Congress rewarded Lt. Newcomb and his crew gold, silver and bronze medals.

The U.S. Life-Saving Service (USLSS) manned coastal observation (surveillance) stations in U.S. waters for the United States Navy. In 1915, Congress combined the USLSS with the USRCS to form the new United States Coast Guard, just in time for the USCG to team up with the U.S. Navy in the domestic and overseas missions of the Great War.[7] The Great War (World War I, 1914–1918) erupted in Europe between the monarchial powers of Europe over colonial competition, territorial disputes, mistrust between the nations that led to militarism and arms races, and the allegedly protective treaty alliances to prevent war that actually fueled the war once the conflict started. The defense treaties forced the nations that signed the mutual defense pacts to enter the war in order to honor their stated prewar obligations. The United States maintained its neutral status until the unrestricted German submarine attacks on merchant shipping drew America into the war on 6 April 1917. A reluctant but pro–British President Woodrow Wilson asked for a declaration of war from Congress and received and signed it. The United States and Britain constituted the nucleus of the Allied Powers. The German Hohenzollern monarchy, Ottoman Turkey, and the Austro-Hungarian Empire constituted the Central Powers alliance.

Before 1917 U.S. Navy warships and U.S. Coast Guard cutters had enforced neutrality laws in American ports and on the Atlantic Ocean. Coast Guard cutters, boats, and shore stations were transferred to the U.S. Navy Department, which received the approximate USCG assets of 220 officers, 47 vessels, 279 Coast Guard stations along the Atlantic and Gulf coasts, and a complement of 4,500 trained and experienced enlisted personnel.

With the coming of the war, the Coast Guard contributed its assets for convoy escort duty on the waters between the United States and Europe, aids to navigation, search and rescue, and port security responsibilities at home and overseas. The port security and SAR responsibilities were assumed on 6 December 1917, when a munitions ship exploded and partially destroyed the Canadian port and town of Halifax, Nova Scotia. More than 10,000 Canadians were killed or injured and up to 3,000 buildings and structures were demolished when the French freighter SS *Mont Blanc*, with a 5,000-ton cargo of explosives, struck the Norwegian ship SS *Imo* in the outer harbor. The *Mount Blanc* had steamed from New York City, one of hundreds of supply ships heading for Europe.

The Coast Guard and its predecessor, the U.S. Revenue Cutter Service, had previously assumed the responsibility of managing foreign and domestic vessels on the lakes, rivers, harbors, bays, and ports along the coasts of the United States. That responsibility continued into World War I. The expertise of the Coast Guard in port security, maritime transportation management, and loading of explosives was extended to and accepted in European ports. The Espionage Act of 1917 extended Coast Guard responsibilities to protect merchant vessels in U.S. waters from sabotage. Waterfront property surveillance and protection, monitoring and regulating vessel movements, guarding restricted and strategic areas, controlling human traffic, and removing people from ships rounded out the port security missions of the USCG. The position of "Captain of the Port" (COTP) was initially established in the busy port of New York and later in other major coastal ports and on the Great Lakes. The COTP had the responsibility of supervising the loading and unloading of explosives from vessels, trucks and trains with trained Coast Guard teams. In World War I, Captain Godfrey L. Carden (USCG), the commander of the New York regional district, was named the COTP of New York City, where most of the munitions shipping occurred. The transportation and personnel statistics from the nearly two-year period of Capt. Carden's command speak for themselves. Coast Guard historian Robert Scheina analyzed the results: "More than 1600 vessels carrying 345 million tons of explosives sailed from the port. In 1918, Carden's division was the largest single command in the Coast Guard, with over 1400 officers and men, four [U.S. Army Corps of] Engineers tugs, and five harbor cutters."[8]

In the fall season of 1917, six cruising (ocean going) U.S. Coast Guard cutters joined U.S. Navy warships in the waters of the East and North Atlantic, and the Mediterranean Sea, as part of the Atlantic Fleet. The cutters escorted hundreds of vessels between the British Isles and the Mediterranean Sea, based out of British Gibraltar on the south coast of Spain. Other cutters patrolled closer to North America from British Bermuda, the Caribbean, and north to Nova Scotia on coastal Canada. The cutters were under the orders of naval district commandants, and the chief of Naval Operations in the Department of the U.S. Navy.

Officers of the U.S. Coast Guard commanded U.S. Navy ships in war zones and were commanders of U.S. Navy training camps. Several Coast Guard officers commanded naval air stations in the United States and Europe, and U.S. Navy transport vessels that carried U.S. troops home from Europe. Seven of the more than 220 commissioned U.S. Coast Guard officers who served in combat zones during the Great War were killed in action.[9] Among those war casualties was the eminent Captain Charles A. Satterlee (USRCS/USCG), who served in the sea service from 1875 to the day of his death in action on 26 September 1918. Third Lt. Satterlee had previously served on the USRC *Levi Woodbury* on blockade and patrol duty in the Spanish-American War of 1898. Prior to 1915 Captain Satterlee served as supervisor of anchorages at Sault Ste. Marie in Michigan, was commander of the USRC *Mackinac,* served on the USRC *Tahoma,* and was assistant inspector of U.S. Life-Saving Service Stations. In 1915 Captain Satterlee was assigned to command the USCGC *Tampa,* the former USRC *Miami.* The *Tampa* would be torpedoed and sunk on 26 September 1918, with the loss off all hands off the southeast coast of the United Kingdom. Two U.S. Navy ships would later be named USS *Satterlee* in honor of Captain Satterlee.[10]

The USCGC *Tampa* was constructed in 1911 as the USRC *Miami* at Newport News (Virginia) Drydock Company and launched on 10 February 1912. The USRC *Miami* was renamed the USCGC *Tampa* on 1 February 1916. The U.S. Navy acquired the *Tampa* on 6 April 1917. The warship displaced 1,181 tons, was 190 feet in length, had a 14-foot draft, 32-foot beam, a speed of 13 knots, and a crew complement of 70 officers and men. In 1917 the *Tampa* listed considerable armament: three 6-pound guns, four 3-inch guns, two machine guns, and two depth-charge racks. The vessel was propelled by one shaft and one 1300 horsepower steam engine. Captain Charles A. Satterlee would be posthumously awarded the Navy Distinguished Service Medal (1918) and, belatedly, in 1999, the Purple Heart.[11] The USCGC *Tampa* had served on peacetime duty out of the port of Tampa, Florida, in the traditional Coast Guard missions of law enforcement, SAR missions, escorting regattas, and doing various festivals. Other *Tampa* missions included the International Ice Patrol out of Halifax, Nova Scotia. When World War I began, the USCGC/USS *Tampa* was assigned to convoy escort duty and antisubmarine warfare (ASW) patrols in European waters, based out of British Gibraltar on the south coast of Spain in the Mediterranean Sea. During its one-year wartime mission, the USS *Tampa* escorted 18 convoys, totaling 350 ships. More that 50 percent of the time the *Tampa* was assigned to European waters the cutter was at sea, patrolling more than 3,500 miles per month. The cutter logs indicate a happy ship, pleasant shoreside recreation, and humanitarian and lifesaving missions performed by the crew and the ship's medical officer.

On the cloudy evening of 26 September 1918, while on patrol in heavy seas in the English Channel off the British coast, crew members on ships in the vicinity of the USS *Tampa* heard a tremendous explosion shortly after the cutter had sailed outbound for what observers assumed was a response to a detected German submarine. The *Tampa* was soon reported missing and did not respond to contact attempts. U.S. destroyers and British patrol vessels searched for the Coast Guard cutter only to find wreckage and two floating bodies wearing naval clothing. No survivors were found. One hundred and eleven Coast Guard crewmembers, four members of the U.S. Navy, eleven Royal Navy seamen, and five civilian passengers were lost. This tragedy that claimed so many lives, from a U-boat attack, meant that the USCG experienced the highest proportional loss of any of the U.S. Armed Forces in World War I. A subsequent naval ceremony on 11 November 1999, at Arlington National Cemetery in Virginia honored the crewmembers of the USS *Tampa*. Given that the USCG was under the jurisdiction of the Department of Transportation in 1999, transportation secretary Rodney E. Slater and Admiral James M. Loy, the Coast Guard commandant, presented the posthumous Purple Heart Medal to the officers and men of the World War I cutter. Accepting the Purple Heart in the name of the USS *Tampa* were the officers and men of the modern USCGC *Tampa* (WMEC 902), in memory of their historic crewmen.[12]

On 5 September 1918, three weeks before the German submarine UB-91 torpedoed the USS *Tampa*, Rear Admiral Albert Parker Niblack (USN), the commander of Squadron 2 Patrol Force, had written a letter to the *Tampa's* commanding officer, Captain Satterlee, commending him and the crew of the cutter for their "excellent record [as] evidence of a high state of efficiency and excellent ship's spirit, and an organization capable of keeping [USS *Tampa*] in service with a minimum of shore assistance." Admiral Niblack added that, as the squadron commander, he took "great pleasure in congratulating [Capt. Satterlee and fellow] officers and crew on the record they have made."[13]

Vice Admiral Sir Charles Holcombe Dare (Royal British Navy) sent a telegram to Admiral William S. Sims (USN) describing the respect and sympathy by the enlisted and officer rates and ranks in the Royal Navy over the tragic loss of the USS *Tampa* and its crew, stating how he and his "staff enjoyed the personal friendship of Captain Charles Satterlee, and had great admiration for his intense enthusiasm and ideals of duty."[14] The British Admiralty sent its collective respect and sympathy in the loss of the USS *Tampa* and its crew: "Their Lordships desire to express their deep regret at the loss of the USS *Tampa* (and the remarkable record of the cutter) since she has been employed in European waters as [part of the] ocean convoy escort 18 convoy from Gibraltar, comprising 350 vessels with the loss of only two ships from enemy action. The convoy commanders recognized the ability with which the *Tampa* carried out the duties of ocean escort. Appreciation of the good work done by the USS *Tampa* may be some consolation to those bereft, and their Lordships would be glad if this could be conveyed to those concerned."[15] In addition to the horrendous Coast Guard crew loss with the sinking of the USS *Tampa*, 81 other USCG personnel died during the Great War because of illness and accidents. A total of 8,835 Coast Guard officers and men served their nation in World War I.[16]

A significant leader in the establishment of U.S Coast Guard and U.S. Navy aviation in the World War I era was Commander Elmer Fowler Stone, who was assigned as a student in April 1916 to U.S. Naval Air Station Pensacola in Florida. Assigned to the U.S. Navy, First. Lt. Stone flew the U.S. Navy NC-4 three-engine floatplane with a USN crew across the Atlantic Ocean in the multistage and first transatlantic flight

Commander Elmer F. Stone (USCG) flew the U.S. Navy floatplane NC-4 across the Atlantic and landed at Lisbon, Portugal, on 27 May 1919.

in 1919. Fourteen U.S. Navy destroyers were strategically stationed across the Atlantic to serve as visual, communication, and potential search and rescue vessels. The flight commenced on 8 May 1919 from New York to Halifax, Nova Scotia, to the insular Azores to Lisbon, Portugal, and then to Plymouth, England, and a harbor landing on 31 May 1919.[17] William H. Thiesen, the Atlantic Area Historian for the U.S. Coast Guard, graphically described the NC-4 flying-boat experience as "exhilarating as well as terrifying," in an aircraft that "had an open unheated cockpit that was very cold because of altitude and wind chill. Tiny glass windshields provided the pilot and co-pilot little protection from on-rushing wind, and three loud Liberty 12 tractor engines, [as] the navigator occupied a cockpit at the exposed forward end of the flying boat's hull."[18]

The exemplary achievements of U.S. Coast Guard ships, boats and personnel at home and abroad, ashore, afloat and in the air demonstrated the skill and significance the Service possessed and performed in domestic security and national defense in partnership with the other U.S. Armed Forces. That experience and record would prepare the Coast Guard to perform those missions across an even broader maritime domain when the Service was called upon to do so again in World War II and subsequent international conflicts.

The famed U.S. Navy seaplane landing at Lisbon, Portugal, in May 1919, flown by then Lt. Elmer Stone (USCG) and his USN crew (United States Coast Guard).

Epilogue

The story of the evolution of the U.S. Revenue Marine, U.S. Revenue Cutter Service, U.S. Coast Guard and U.S. Navy warships, from the age of sail- and wind-powered vessels to the Civil War ironclads and iron-hulled ships, a submarine, and auxiliary (steam and sail) wooden and iron vessels is a compelling story. The power and steering sources evolved from the stern sailing rudder to a helm steering wheel amidships attached to a stern rudder; steam-powered paddle-wheel ships; and finally, the modern propeller, or screw, attached to shafts connected to coal- and then oil-burning machinery in the engine room below the main deck. David O. Whitten summarized naval historian Professor Craig Symonds' classic *U.S. Navy History*. Symonds chronicled the evolution of naval vessels from the nineteenth through the twenty-first centuries and the transition from wooden- to steel-hulled surface vessels and boats powered by coal and steam, gasoline, petroleum, and eventually nuclear fuel. The mechanical physics and engineering complexities involved in that evolution challenged naval and nautical engineering experts and the personnel who trained the naval crews in their increasingly complex responsibilities.[1]

The sailing, or wind-powered, era spanning several centuries is perhaps the most enduring and challenging era. Peter Stanford wrote in fascinating detail about sailing ships with lengths of several hundred feet and masts and sail heights of 60 to 100 feet or more above the deck, requiring the agility and courage of seamen with their skilled and dangerous tasks to climb the rigging and manage the sails. The one-, two- and three-masted vessels were nautically and structurally arranged to maximize efficiency and speed in a variety of seas and winds and in calm and stormy waters. The ship construction scheme and terminology from bow to stern included the fore bowsprit to attach sail lines to, and the foremast, mainmast, and rear mizzenmast. The ascending sails on the foremast were termed the foretopmast, and then the foretopgallant mast. The terminology dates from Renaissance, British and American naval and maritime history. The center, or mainmast, from deck to top included the mainsail, then the lower topsail, upper topsail, topgallant, and royal sail and sky sail. The third mast, or stern mast, was termed the mizzen. The Renaissance term "topgallant" was pronounced "t'gallant," and referred to being "extra" or "over the top."[2]

Among the fascinating and heroic stories from the sailing ship era in the War of 1812 between the United States and the United Kingdom were the naval battles against the Royal Navy and His Majesty's warships by American naval forces on the Great Lakes, as well as on the ocean and Gulf Coast, Chesapeake Bay, and Mississippi River. Vice Admiral Peter H. Daly, USN (Ret.), offered an account of the Battle of Lake Erie in the 4 July 2016 newsletter of the U.S. Naval Institute on the anniversary of the July day in 1813 when the U.S. brig *Niagara* was launched into an adjacent bay to join Commodore Oliver Hazard Perry and his squadron. In that year, the fledgling "United States was fighting for its existence"[3] in what historians have termed America's "Second War of Independence" against Britain. Daly described the 10 September naval clash on Lake Erie between Perry aboard his flagship, USS *Lawrence*, against Captain R.H. Barclay (Royal Navy). In the heat of battle, the *Lawrence* lost its cannon fire and maneuverability, forcing Perry to embark on a rowboat voyage to the *Niagara*. Daly quotes Theodore Roosevelt, the future U.S. president, from Roosevelt's acclaimed book titled *The Naval War of 1812*: "The British ships fought themselves to a standstill … the *Niagara* … within pistol-shot, keeping up a terrific discharge of great guns and musketry … and Barclay's flag was struck … [giving the United States naval forces] complete command of all the upper lake [which] prevented invasion from that quarter [and] increased our prestige with the foe."[4]

A historic legacy and commemoration of the Age of Sailing Ships for the United States Coast Guard is the USCGC *Eagle* (WIX-327), the training ship used at the U.S. Coast Guard Academy in New London, Connecticut. The 295-foot cutter was constructed and launched in Germany in 1936, three years before the outbreak of World War II. Vice Admiral Paul Welling (USCG, Ret.) provided an extensive description of the history and mission of the *Eagle*. The following is based on Admiral Welling's scholarship. The *Eagle* arrived at the U.S. Coast Guard Academy in 1947, two years after the end of World War II. The training cutter is used today to give traditional sailing ship experience and seafaring knowledge to USCGA cadets and commissioned officers. The tall ship is a three-mast barque with twenty-three sails, "miles of line, halyards, sheets, clew lines, and other rigging leading up to the masts and yards, braces and tacks running fore and aft, with square sails furled overhead. [Academy cadets will be] climbing the shrouds to the top and out on the footropes slung below the yards."[5]

The USCGC *Eagle* (WIX 327) is the wind-powered sailing cadet-training ship for future officers at the United States Coast Guard Academy in New London, Connecticut. It was acquired from the German navy as a war prize for reparations by the United States government after World War II. In 1946 a mixed Coast Guard and German crew sailed the former *Horst Wessel* across the Atlantic to New London, Connecticut. The 295-foot "tall ship" is a cutter of full rigging and masts, and cadets must learn to scale the rigging. The mainmast is 147 feet off the deck. The cutter holds a 1,000-horsepower diesel engine, and the crew complement numbers 8 commissioned officers, 50 assigned crewmembers, and 150 cadet trainees and officer candidates on any given voyage. The *Eagle* sails international waters and represents the U.S. Coast Guard and the United States government in domestic and foreign ports.

The U.S. Coast Guard, while fighting in combat in maritime theater during the Second World War, acquired the German barque *Horst Wessel* at Kiel, Germany, as part of mandated war reparations in 1945. The German training ship was renamed the *Eagle*, and in June of 1946 commenced its long voyage across the Atlantic to the United States under the command of Captain Gordon McGowan (USCG) and a crew complement of Coast Guard and German sailors. The skilled seamen guided the cutter through a hurricane, and the sailing cutter's shredded sails were visible to the public when the ship docked at New York City.[6]

On the *Eagle* cruises in U.S. and international waters, the crew consists of cadets, commissioned officers, enlisted experienced Coast Guard personnel, and a total crew complement of approximately 50, to instruct the trainees on board during each training cruise. The *Eagle* sails domestic and international waters to train its future maritime leaders and participate in international sailing ship events and ceremonies that draw spectators in the hundreds and thousands.[7] The training that the *Eagle* crew members receive on the wind-powered sailing cutter prepares the men and women for their future assignments in the multi-mission Coast Guard.

These future Coast Guard leaders gain an appreciation of the heritage of naval ships and the traditions that have provided the modern U.S. Navy and U.S. Coast Guard with the legacies their own contributions will enhance. That invaluable training prepares Coast Guard members to live up to the Service motto:

Semper Paratus, "Always Ready."

Appendix: Maps

The following maps (by David H. Allen) demonstrate and clarify the complex historical geography of U.S. Revenue Cutter and U.S. Coast Guard missions from the origins of the Service in 1790 through the Great War (World War I) of 1914–1918. The maritime domain covered by the U.S. Revenue Marine/U.S. Revenue Cutter Service/U.S. Coast Guard in domestic waters and overseas in its joint missions with the U.S. Navy is extraordinary for that relatively small sea service.

In chronological and geographical order the maps illustrate terrestrial and maritime locations and sites in the Quasi-War with France (1798–1801); the War of 1812 against Britain in what some historians have called America's Second War of Independence; the maritime theater of operations in the War of 1812 in the vast interior waterways of the Great Lakes in the northern part of the United States; and Chesapeake Bay along the East Coast in proximity to the Atlantic Ocean.

Between 1800 and 1850 the USRM and USN were involved in the interdiction of slave ships and pirate ships in the Atlantic and in the Gulf of Mexico and the Caribbean Sea. Naval operations shifted to the Gulf of Mexico and the coastal waters of Mexico in the War with Mexico (1846–1848). Then the maritime theaters of operations shifted to the Atlantic Coast, the Gulf, interior waterways, and the Mississippi River in the Civil War (1861–1865).

The Spanish-American War of 1898 involved U.S. military and naval forces and joint operations on and off the shores of Cuba in the Gulf and Caribbean regions, and the western Pacific in the Philippine islands.

The Great War (1914–1918), later called World War I, required joint U.S. Navy and U.S. Coast Guard operations along the Atlantic Coast in port security antisubmarine warfare and convoy escort missions across the Atlantic and into the Mediterranean Sea, and north to the waters off northwest Europe and the United Kingdom.

It is hoped the maps will facilitate a better understanding for the reader regarding the range, scope and complexities of the joint national defense role in the history of the United States performed by the U.S. Navy and U.S. Coast Guard from the inception of the sea services in the late 18th century into the early part of the 20th century.

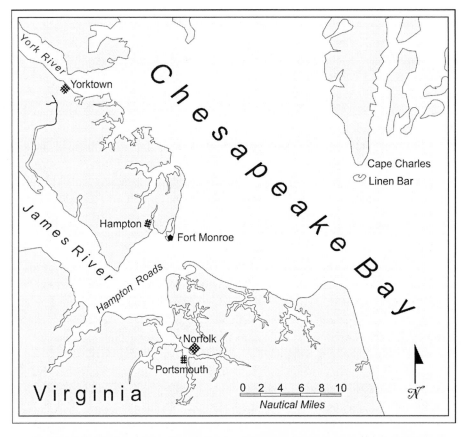

The Potomac River flows southeast from Washington, D.C., and into the Chesapeake Bay, which leads into the Atlantic Ocean. That interior and coastal maritime realm was a significant theater of action in the Civil War. Land and sea combat operations occurred in Virginia at Yorktown on the York River, at Hampton, Hampton Roads, and Fort Monroe by the James River, and at Norfolk and Portsmouth. These sites saw military and naval action between Confederate and Union troops and the U.S. Navy and Marines and the U.S. Revenue Cutter Service.

The War of 1812 is also called the Second War of Independence with Britain. The maritime combat areas included the Great Lakes, the Atlantic and Gulf coasts, the Chesapeake Bay, and the Mississippi River. The Revenue Cutter *Vigilant* fought and captured the British vessel *Dart*. RC *Eagle* battled the HMS *Dart*. The Great Lakes (Ontario and Erie) and Lake Champlain were significant maritime battle areas, as were the Gulf and Atlantic coasts, for U.S. Navy and U.S. Revenue Marine combat action. Other cutters included the *Jefferson, Gallatin, Surveyor,* and *Madison.*

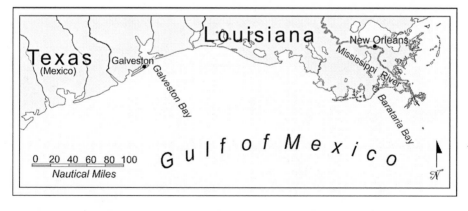

Cutter crews and the U.S. Navy and Royal Navy confronted pirates and slave ships in their anti-pirate and anti-slave missions in the Caribbean and Atlantic. Slaves were released. Pirates were battled and captured in the Mediterranean, the Gulf of Mexico, and along the lower Mississippi River.

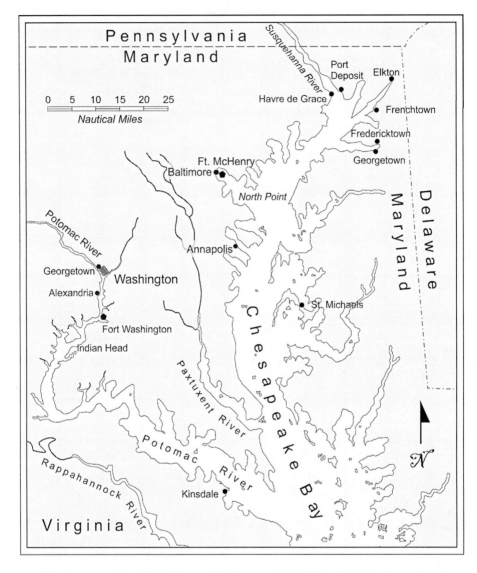

The interior waters of the Chesapeake Bay were a significant theater of operations in the war against the Royal Navy, privateers, and British soldiers and marines in the War of 1812. The U.S. Revenue Marine engaged in anti-smuggling, espionage, transportation, and reconnaissance missions, and carried dispatches, troops and supplies. U.S. Navy and Revenue Marine vessels clashed with British barges, boats and warships in all maritime theaters, including the waters off British Canada.

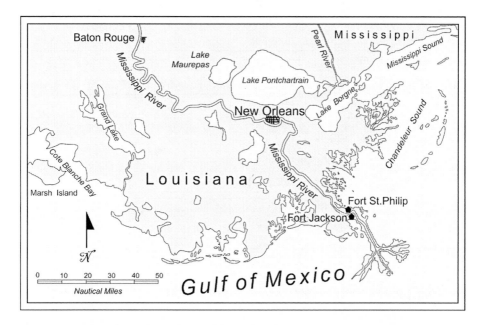

Revenue cutters participated with the U.S. Navy, U.S. Marines, and U.S. Army in every theater of war during the Civil War, including during the ascent up the Mississippi River to New Orleans with Commodore David Farragut (USN) in the flagship USS *Hartford* and his naval squadron. The USRC *Harriet Lane*, serving as a U.S. Navy warship, joined the mission and served gallantly.

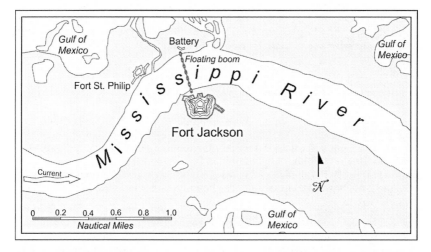

The ascent up the Mississippi to New Orleans in the Civil War required passing through the formidable Confederate fortress defenses, cannons, and barrier of the floating boom between Fort St. Philip and Fort Jackson at Plaquimine Bend in a nighttime assault that involved fire rafts and the firing of enemy guns upon the wooden sail and steam ships of the U.S. fleet. As the Confederate batteries were being suppressed by naval warships and gunboats, the USS *Hartford* proceeded north to capture the Southern port of New Orleans.

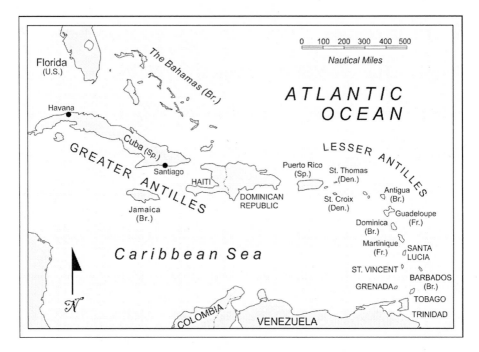

The cutters battled pirate ships and crews in the maritime realms of Louisiana and northern Latin America. The Revenue Cutters and crews of the *Louisiana* and the *Alabama* were mission leaders. The U.S. Revenue Marine, U.S. Navy, U.S. Marines and U.S. Army also battled the brave Seminole Indians in the Seminole Wars between 1820 and 1850 in the interior swamps and coastal regions of Florida.

Opposite, top: Eight sail/wind-powered revenue cutters operated along the South Atlantic coast and in the West Indies and Caribbean Sea, carrying from 10 to 14 guns and crew complements of 30 to 70 officers and men. The cutters captured 18 French vessels and aided the U.S. Navy in other captures. The Revenue Cutter *Pickering* was a particularly successful cutter.

Opposite, bottom: The U.S. Navy and U.S. Revenue Cutter Service fought Cuban shore batteries and the Spanish Fleet in the vastness of the Atlantic Ocean and Caribbean Sea. Illustrated here is the linear extent of the island of Cuba, 90 miles south of Florida and under the control of Spain, and Spanish Puerto Rico. The logistical and tactical operations of the U.S. Navy and U.S. Revenue Cutter Services were considerable and complex in that extensive maritime domain. The Revenue Cutter Service and its corollary agencies, the U.S. Life-Saving Service and U.S. Lighthouse Service, also served surveillance units to protect against the potential presence of Spanish naval and military activities in U.S. waters.

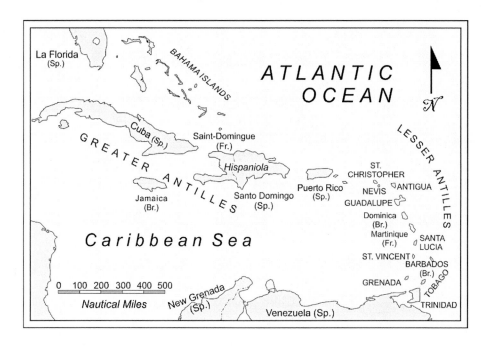

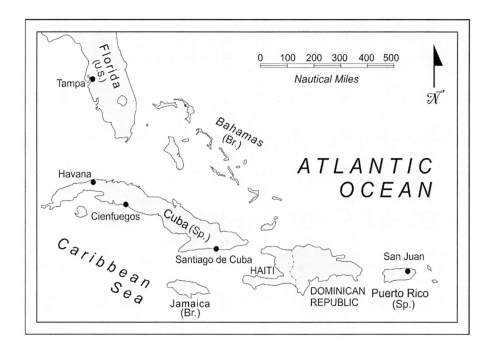

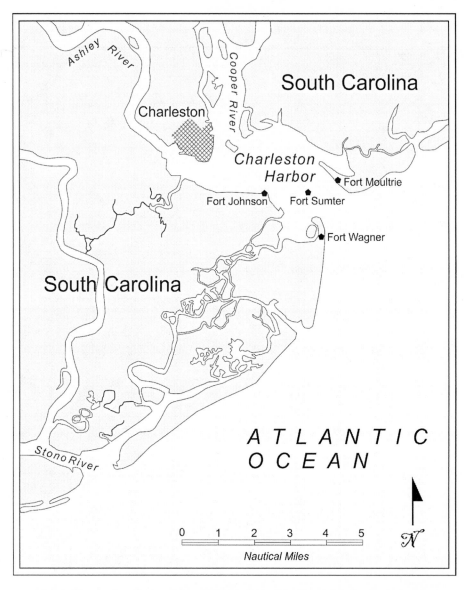

The Civil War erupted on April 12, 1861, with the firing by Confederate troops on the Union's Fort Sumter in Charleston Harbor. Fort Sumter was under the command of Major Robert Anderson. Union naval forces failed to recover Ft. Sumter until military and naval forces completed the recapture near the end of the war. During the war, the U.S. Revenue Cutter Service supported the U.S. Navy blockade of the Confederate ports along major rivers and the Atlantic and Gulf coasts. The USRC *Harriet Lane* fired the first naval shots of the war in Charleston Harbor at the beginning of the war. The USRC *Miami* transported President Abraham Lincoln and other federal officials to Fort Monroe in Virginia prior to the Peninsular Campaign. Cutters enforced the naval blockade, chased Confederate commerce raiders, transported troops, supplies, and government documents, and got troops ashore in amphibious assaults.

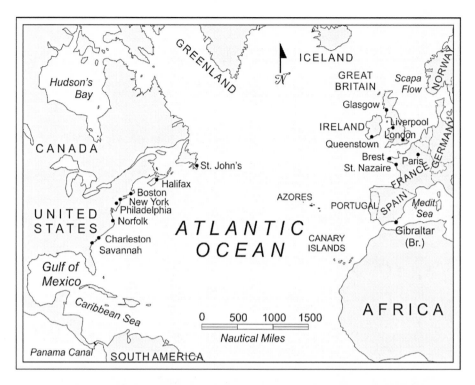

The U.S. Life-Saving Service was combined with the U.S. Revenue Cutter Service to form the United States Coast Guard in 1915. World War I was fought from 1914 to 1918, but the United States did not enter the war until 1917. The USN and USRCS, now the USCG, performed convoy escort service and antisubmarine warfare patrols across the Atlantic and between the Mediterranean Sea, the eastern Atlantic, France and the United Kingdom. The USCG served under the command of the U.S. Navy in domestic waters and overseas in port security, combat, transport, and dangerous cargo loading missions. Cutters and personnel were stationed in European ports and at the British port of Gibraltar in the Mediterranean. USCG officers and crews ran cutters and were part of USN ships and crews. The U.S. Coast Guard cutters *Algonquin, Manning, Ossippee, Seneca, Yamacraw* and *Tampa* sailed with the U.S. Navy. Coast Guard officers commanded U.S. Navy units. After acclaimed convoy escort service, the heavily armed 190-foot USCGC *Tampa* (under the command of Capt. Charles Satterlee, USCG), was sunk by a German submarine off the coast of Wales with the loss of over 100 crewmen. The British Admiralty paid tribute to the *USCG* and the *Tampa* for exemplary seamanship and service.

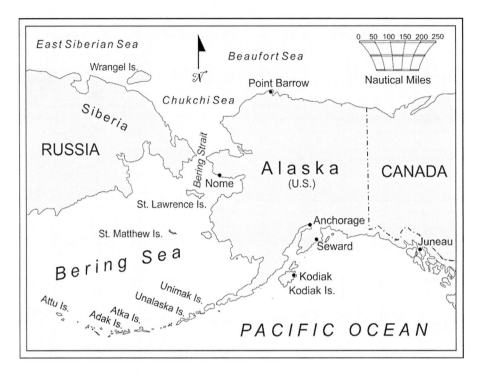

The United States purchased what was then Russian Alaska in 1867 as a result of the diplomatic negotiations of Secretary of State William Seward. The USRCS was assigned the duties of federal law enforcement, search and rescue, environmental and fauna protection, and national defense in that vast terrestrial and maritime realm. The USRCS transported government officials and supplied the Inuit and Indian populations as well as providing a platform for judicial proceedings and scientific research. Cutters that patrolled this treacherous environment included the *Lincoln, Reliance, Corwin, Bear* and *Wyanda,* manned by their extraordinary, experienced line officers, medical officers, commanders and crews. The region is considered in this book to illustrate the tremendous geographical range and heterogeneous duties and responsibilities performed by the USRCS and USCG as the sea service carried out its multiple missions at home and overseas.

Opposite: The USRCS supported the U.S. Navy in combat, escort, blockade communications, and search and rescue missions off the coast of Cuba, in the Gulf of Mexico and Caribbean, and in the Philippines campaign in the Pacific. The USRCS served under Rear Admiral William T. Sampson (USN) off the coast of Cuba in blockade duty. The RC *Manning* fired on Spanish shore batteries. Eight revenue cutters carrying 43 guns assisted USN vessels in the Cuban campaign. Under fire and firing back at Spanish guns, the RC *Hudson* towed the battle-damaged USS *Winslow* and its heavy crew casualties out of harm's way.

The USRC *McCulloch* carried 100 officers and men and six guns under the command of Commodore George Dewey (USN) in combat, dispatch and supply-ship protection missions between Hong Kong and Manila Bay.

Chapter Notes

Introduction

1. John W. Chambers II, editor in chief, *The Oxford Companion to American Military History* (New York: Oxford University Press 1999), 145.

2. Robert Scheina, "The Coast Guard at War," U.S. Department of Homeland Security, United States Coast Guard, Office of the U.S. Coast Guard Historian. http://www.uscg.mil/history/articles/h_CGatwar.asp (accessed 23 October 2014).

3. *Ibid.*

4. *Ibid.*

5. *Ibid.*

6. Thomas P. Ostrom, *The United States Coast Guard, 1790 to the Present* (Oakland, OR: Red Anvil/Elderberry Press, 2006), 11.

7. *Ibid.*, 11, 16.

8. John J. Galluzzo, "1790–1915: That Others Might Live," Tom Beard, editor-in-chief, *The Coast Guard* (Seattle, WA: Foundation for Coast Guard History, and Hugh Lauter Levin Associates, 2004, 33, 35, 41. Distributed by Publishers Group West. Tom Beard, editor-in-chief, *The Coast Guard.*

9. *Ibid.*, 43–45, 55.

10. Scheina, "Coast Guard at War."

11. *Ibid.*

12. *Ibid.*

13. *Ibid.*

14. Stephen H. Evans, *The United States Coast Guard, 1790–1915* (Annapolis, MD: The United States Naval Institute, 1949), 6–7.

15. *Ibid.*, 92–96,151, 155, 157, 198–199.

16. Scheina, "Coast Guard at War."

17. *Ibid.*

18. *Ibid.*

19. Ostrom, *The United States Coast Guard, 1790 to the Present*, 42.

20. Evans, 3, 18.

21. "Ellsworth P. Bertholf," U.S. Department of Homeland Security, and United States Coast Guard, Office of the Coast Guard Historian. http://www.uscg.mil/history/people/EP BertholfBio.asp (accessed 25 October 2014).

Chapter 1

1. "Eighteenth, Nineteenth, & Early Twentieth Century Revenue Cutters." U.S. Department of Homeland Security and United States Coast Guard. U.S. Coast Guard Historian's Office. http://www.uscg.mil/history/webcutters/ USRC_Photo_Index.asp (accessed October 29, 2014).

2. "The First Ten Cutters." U.S. Department of Homeland Security and United States Coast Guard. U.S. Coast Guard Historian's Office. http://www.uscg.mil/history/webcutters/ First_Cutters.asp (accessed October 29, 2014).

3. Howard I. Chapelle, *The History of American Sailing Ships* (New York: W.W. Norton & Co., 1935), 176.

4. *Ibid.*

5. *Ibid.*, 177–181.

6. *Ibid.*, 181–182.

7. *Ibid.*, 184–201.

8. *Ibid.*, 200–204.

9. Spencer C. Tucker, "Armaments & Innovations: The U.S. Navy's 'Smashers,'" *Naval History*, December 2014, 10–12.

10. Chapelle, *American Sailing Ships*, 204–210.

11. *Ibid.*, 210.

12. *Ibid.*, 210–211.

13. *Ibid.*, 212–214.

14. *Ibid.*, 214–216.

15. *Ibid.*, 216–218.

16. Donald L. Canney, *U.S. Coast Guard and*

Revenue Cutters, 1790–1935 (Annapolis, MD: Naval Institute Press, 1995), xiii, xv–xvi.

17. *Ibid.*, xvii.

18. *Ibid.*, 1–15.

19. *Ibid.*, 27–28, 35–36, 40–42, 45.

20. Thomas P. Ostrom, *The United States Coast Guard on the Great Lakes* (Oakland, OR: Red Anvil/Elderberry Press, 2007), 55–56.

21. Canney, *U.S. Coast Guard and Revenue Cutters*, 46.

22. Ostrom, *The U.S. Coast Guard on the Great Lakes*, 56–57.

23. *Ibid.*, 16–17.

24. Canney, 47–52.

25. *Ibid.*, 53–61, 66–70.

26. Dennis L. Noble, *That Others Might Live: The U.S. Life-Saving Service, 1878–1915* (Annapolis, MD: Naval Institute Press, 1994), 27.

27. *Ibid.*, 27–33.

28. *Ibid.*, 20, 23, 45, 107, 150–155.

29. Chapelle, *American Sailing Ships*, 218.

Chapter 2

1. Craig L. Symonds, *Historical Atlas of the U.S. Navy* (Annapolis, MD: Naval Institute Press, 1995), 26–27.

2. *Ibid.*, 29.

3. Robert Scheina, "The Coast Guard At War," U.S. Department of Homeland Security, United States Coast Guard, U.S. Coast Guard Historian's Office. http://www.uscg.mil/history/articles/h_CGatwar.asp (accessed 23 October 2014).

4. John W. Chambers II, editor-in-chief, *The Oxford Companion to American Military History* (New York: Oxford University Press, 1999), 280–281.

5. William H. Thiesen, "Benjamin Hiller & the Cutter *Pickering* in the Quasi-War with France," *Sea History*, Spring 2008, 24–27.

6. *Ibid.*, 24–25.

7. *Ibid.*, 25.

8. *Ibid.*, 26.

9. *Ibid.*, 27.

10. *Ibid.*

11. *Ibid.*, 26.

12. *Ibid.*, 24.

13. Chambers, *Oxford Companion to American Military History,* 280–281.

Chapter 3

1. George Brown Tindall, and David E. Shi, *America: A Narrative History* (New York: W. W. Norton & Co., 1996), 376–380.

2. Samuel Eliot Morison, Henry Steele Commager, and William Leuchtenburg, *The Growth of the American Republic*, Vol. I (New York: Oxford University Press, 1969), 357–365.

3. Samuel Eliot Morison, *The Oxford History of the American People* (New York: Oxford University Press, 1965), 390–391.

4. *Ibid.*, 391–397.

5. Morison, Commager, Leuchtenburg, *Growth of the American Republic*, 383–386.

6. Larry Schweikart and Michael Allen, *A Patriot's History of the United States* (New York: Sentinel/Penguin Group, 2007), 175.

7. *Ibid.*, 176.

8. Morison, *Oxford History of the American People*, 398–399.

9. Schweikart and Allen, *A Patriot's History*, 177.

10. Walter R. Borneman, *1812: The War That Forged A Nation* (New York: Harper Collins, 2005), 1–3.

11. *Ibid.*, 298.

12. *Ibid.*, 304.

13. "When the U.S. Invaded Canada," *The Week*, August 10, 2012, 11.

14. *Ibid.*

15. Ray Allen Billington and Martin Ridge, *Westward Expansion: A History of the American Frontier* (New York: Macmillan Publishing Co., 1982).

16. *Ibid.*, 267–269.

17. *Ibid.*, 269–271.

18. *Ibid.*, 271.

19. *Ibid.*, 271–272.

Chapter 4

1. Stephen H. Evans, *The United States Coast Guard, 1790–1915: A Definitive History* (Annapolis, MD: The United States Naval Institute, 1949), 18–19.

2. *Ibid.*, 19.

3. *Ibid.*, 20–21.

4. *Ibid.*, 21–22.

5. John Whiteclay Chambers II (editor-in-chief), *The Oxford Companion to American Military History* (New York: Oxford University Press, 784), 1999.

6. Mark Lardas, *Great Lakes Warships, 1812–1815* (Long Island City, NY: Osprey, 2012), 4–7.

7. *Ibid.*, 45–46.

8. *Ibid.*, 8–9.

9. Craig L. Symonds, *Historical Atlas of the U.S. Navy* (Annapolis, MD: Naval Institute Press, 1995), 48–49.

10. *Ibid.*

11. Craig L. Symonds, *Historical Atlas of the U.S. Navy* (Annapolis, MD: Naval Institute Press, 1995), 50–51. The geographic summations described in Symonds's U.S. Navy histories throughout this book are based on this author's descriptions of the magnificent maps in Symonds's *Historical Atlas of the U.S. Navy*, as constructed and drawn by the former U.S. Naval Academy and later freelance cartographer William J. Clipson. Professor Craig Symonds was professor of naval history and Civil War history at the U.S. Naval Academy. Symonds also taught at the U.S. Naval War College in Newport, Rhode Island.

12. *Ibid.*

Chapter 5

1. William H. These, "Captain Frederick Lee and Cutter Eagle in the War of 1812," History Lesson Bulletin, by The Foundation for Coast Guard History and published in *The Bulletin*, the magazine of the U.S. Coast Guard Academy Alumni Association, June 2012, 60.

2. *Ibid.*, 56.

3. *Ibid.*, 61.

4. *Ibid.*

5. Paul H. Johnson, "The Search for Captain Frederick Lee," *The Bulletin*, March/April 1977, 34–36.

6. *Ibid.*, 34–35.

7. *Ibid.*, 35.

8. *Ibid.*, 35–36.

9. *Ibid.*, 36.

10. *Ibid.*

11. "Loss of the U.S. Revenue Schooner *Gallatin*, 1 April 1813 in Charleston Harbor, South Carolina," United States Coast Guard History Program, United States Coat Guard, U.S. Department of Homeland Security, Coast Guard Historian's Office, 1 (accessed November 16, 2014). The excerpted transcript is transcribed from the 2 April 1813 edition of the *Charleston Courier*, as quoted from the *Early History of the United States Revenue Marine Service* (United States Revenue Cutter Service), 1789–1849), by Horatio Davis Smith. The source was edited by Elliott Snow (Naval Historical Foundation, and the Press of R.L. Polk Printing Co., 1932); and reprinted by the U.S. Coast Guard, 1989.

12. *Ibid.*, 1–2.

13. *Ibid.*, 2.

14. "Great Lakes Revenue Vessels," United States Coast Guard and the U.S. Department

of Homeland Security, U.S. Coast Guard History Program, 1 (accessed December 17, 2014).

15. *Ibid.*

16. *Ibid.*

17. *Ibid.*

18. *Ibid.*, 2.

19. *Ibid.*, 3.

20. *Ibid.*

21. "War of 1812 Revenue Cutter and Naval Operations," United States Coast Guard and U.S. Department of National Security, U.S. Coast Guard History Program, U.S. Coast Guard Historian's Office, Washington, D.C. (page numbers are not cited in the 22 page document). Accessed December 15, 2014. The document is arranged chronologically by Revenue Marine incidents and missions: April 4–December 29, 1812; January 17–December 17, 1813; January 1–December 24, 1814; and January 4–July 4, 1815. The author has omitted events chronicled in the document that have been mentioned elsewhere in this book.

22. *Ibid.*

23. *Ibid.*

24. *Ibid.*

25. *Ibid.*

26. *Ibid.*

27. *Ibid.*

28. *Ibid.*

29. *Ibid.*

30. *Ibid.*

Chapter 6

1. William H. Thiesen is the Atlantic Area Historian of the United States Coast Guard. Prior to that assignment, Thiesen served as assistant director and curator of the Wisconsin Maritime Museum. He has authored numerous articles and contributed to several periodicals. Dr. Thiesen is the author of several books, including *Industrializing American Shipbuilding: The Transformation of Ship Design and Construction, 1810–1920* (2006), and *Cruise of the Dashing Wave: Rounding Cape Horn in 1860* (2010).

2. William H. Thiesen, "They Did Their Duty as Became American Sailors," *Naval History*, August 2014, 52.

3. *Ibid.*, 52–57.

4. *Ibid.*, 53.

5. *Ibid.*

6. *Ibid.*, 54–57.

7. William H. Thiesen, "With a 'Spirit and Promptitude,'" *Proceedings*, August 2012, 30.

8. *Ibid.*, 31.

9. *Ibid.*, 32.

10. *Ibid.*

11. William H. Thiesen, "United States Coast Guard, War of 1812," Revenue Cutter Operations and the Core Coast Guard Missions, U.S. Coast Guard Historian's Office, U.S. Coast Guard Headquarters, Washington, D.C., 1–15 (accessed January 4, 2015).

12. *Ibid.*, 2.

13. *Ibid.*, 2–3.

14. *Ibid.*, 3.

15. *Ibid.*, 3–4.

16. *Ibid.*, 4–5.

17. *Ibid.*, 5.

18. *Ibid.*, 5–6.

19. *Ibid.*, 11.

20. *Ibid.*, 8, 12.

21. *Ibid.*, 13.

22. *Ibid.*, 15.

23. Craig L. Symonds, *Decision at Sea: Five Naval Battles That Shaped American History* (Oxford, NY: Oxford University Press, 2005), 4, 5.

24. William H. Thiesen, "United States Coast Guard, War of 1812," 15.

Chapter 7

1. Craig L. Symonds, *Historical Atlas of the U.S. Navy* (Annapolis, MD: Naval Institute Press, 1995), 62–63.

2. Stephen H. Evans, *The United States Coast Guard, 1790–1915: A Definitive History* (Annapolis, MD: The United States Naval Institute, 1949), 22–23.

3. *Ibid.*, 23.

4. *Ibid.*, 23–24.

5. *Ibid.*, 25.

6. *Ibid.*

7. *Ibid.*, 25–26.

8. *Ibid.*, 24–26.

9. *Ibid.*, 26.

10. *Ibid.*

11. "(RC) *Louisiana, 1804*," U.S. Coast Guard, Coast Guard Historian's Office, https://www.uscg.mil/history/.../Louisiana.pdf (accessed December 20, 2014). This RC *Louisiana* was commissioned in 1804 and decommissioned in 1812. This cutter preceded the same named RC *Louisiana* commissioned in 1819, and sold in 1833 that also engaged in the piracy wars.

12. Robert Scheina, "Seminole Wars," The Coast Guard At War: A History, U.S. Department of Homeland Security and United States Coast Guard, http://www.uscg.mil/history/articles/h_CGatwar.asp (accessed January 22, 2015).

13. "U.S. Coast Guard History: Historical Overview," U.S. Department of Homeland Security and United States Coast Guard, Coast Guard Historian's Office, http://www.uscg.mil/history/articles/h_USCGhistory.asp (accessed December 20, 2014).

14. *Ibid.*

15. *Ibid.*

16. *Ibid.*

17. Stuart Murray, *Atlas of American History* (New York: Checkmark Books, 2005), 58–59.

18. John Whiteclay Chambers II (editor-in-chief), et. al., *The Oxford Companion to American Military History* (New York: Oxford University Press, 1999), 646–647.

19. *Ibid.*

Chapter 8

1. Robert Scheina, "The Coast Guard at War," U.S. Department of Homeland Security and United States Coast Guard, http://www.uscg.mil/history/articles/h_CGatwar.asp (accessed 23 October 2015).

2. *Ibid.*

3. John Whiteclay Chambers II (Editor-in-Chief), et. al, *The Oxford Companion to American Military History* (New York: Oxford University Press, 1999), 433.

4. *Ibid.*

5. *Ibid.*

6. *Ibid.*, 434–435.

7. Stuart Murray, *Atlas of American Military History* (New York: Checkmark Books, 2005), 62–67.

8. Craig L. Symonds, *The Naval Institute Historical Atlas of the U.S. Navy* (Annapolis, MD: Naval Institute Press, 1995), 68–69.

9. *Ibid.*, 70–71.

10. *Ibid.*

11. Robert Ayer, editor, and Gary M. Thomas, executive director of the Foundation for Coast Guard History, "Mexican War: 1846–1848," *The Cutter Newsletter*, Summer/Fall 2014, 20.

12. *Ibid.*

13. *Ibid.*

14. *Ibid.*

15. Stephen H. Evans, *The United States Coast Guard, 1790–1915: A Definitive History*, Annapolis, MD: The United States Naval Institute, 1949), 38, 50.

16. *Ibid.*

17. *Ibid.*, 59–60.

18. *Ibid.*, 60.

19. *Ibid.*

20. *Ibid.*, 60–61.
21. *Ibid.*, 61.
22. *Ibid.*
23. *Ibid.*, 61–63.
24. *Ibid.*, 63.
25. *Ibid.*

Chapter 9

1. James M. McPherson, *Embattled Rebel: Jefferson Davis as Commander in Chief* (New York: Penguin Press, 2014), 60, 82, 107.
2. Walter Star, *Seward: Lincoln's Indispensable Man* (New York: Simon & Schuster, 2012), 83–84, 230–233, 236–237. 249–251, 290–292, 330, 341–347, 462, 482–491.
3. *Ibid.*, 1–4.
4. Thomas R. Turner, *The Civil War: 101 Things You Didn't Know About It* (Avon, MA: Adams Media, 2007), xi, xii, 1–9.
5. Kenneth M. Stamp (contributor and editor), *The Causes of the Civil War* (New York: Simon & Schuster, 1991. The page numbers that introduce each philosophical or causal section of the book are 13, 19, 59, 85, 107, 135, 185, and 201.
6. Bruce Catton, *The Civil War* (Boston, MA: Houghton Mifflin Co., 2004), 7–21.
7. *Ibid.*, 22–24.
8. *Ibid.*, 25–26.
9. *Ibid.*, 27–33.
10. *Ibid.*, 68–75.

Chapter 10

1. John Whiteclay Chambers II (editor), *The Oxford Companion to American Military History* (New York: Oxford University Press, 1999), 176.
2. *Ibid.*
3. *Ibid.*, 176–177.
4. Tom Chaffin, *Sea of Gray: The Confederate Raider Shenandoah* (New York: Hill and Wang, 2006), 15–16, 30–31, 37, 49–51, 72–74, 226–228, 246, 318, 376.
5. Mark Lardas, *CSS Alabama vs USS Kearsarge: Cherbourg, 1864* (Oxford, United Kingdom, and Long Island City, NY: Osprey, 2011), 4–7, 32, 55–63.
6. Patricia L. Faust, editor, *Historical Times Illustrated Encyclopedia of the Civil War* (New York: Harper & Row, 1986, 3–4, 409.
7. *Ibid.*, 666–667.
8. *Ibid.*, 836.
9. Chambers, *The Oxford Companion to American Military History*, 742–743.

10. Faust, *Encyclopedia of the Civil War*, 813.
11. *Ibid.*, 283.
12. *Ibid.*, 254.
13. *Ibid.*, 86, 504.
14. *Ibid.*, 594.
15. *Ibid.*, 281–282.
16. Benton Rain Patterson, *The Mississippi River Campaign, 1861–1863: The Struggle for Control of the Western Waters* (Jefferson, NC: McFarland, 2010), 106, 118.
17. *Ibid.*, 119.
18. Chambers, *The Oxford Companion to American Military History*, 201.
19. Stuart Murray, *Atlas of American History,* New York: Checkmark Books, 2005), 82–83.
20. *Ibid.*, 100–101.
21. *Ibid.*, 96.
22. *Ibid.*, 96–97.
23. Craig L. Symonds, *The Naval Institute Historical Atlas of the U.S. Navy* (Annapolis, MD: Naval Institute Press, 1995), 76–77, 78, 82, 83.
24. *Ibid.*, 84–85.
25. *Ibid.*, 94–95.
26. *Ibid.*, 96.
27. *Ibid.*, 96–97.

Chapter 11

1. Patricia L. Faust, editor, *Historical Times Illustrated Encyclopedia of the Civil War* (New York: Harper & Row, 1986), 624–625.
2. *Ibid.*, 342, 344.
3. Benton Rain Patterson, *The Mississippi River Campaign, 1861–1863* (Jefferson, NC: McFarland, 2010), 103, 118.
4. Kevin J. Dougherty, *Ships of the Civil War, 1861–1865* (New York: Metro Books/Sterling Publishing Co., Inc./Amber Books, Ltd., 2013), 92–93.
5. Stephen H. Evans, *The United States Coast Guard, 1790–1915* (Annapolis, MD: United States Naval Institute, 1949), 73–75.
6. *Ibid.*, 78–85.
7. Robert M. Browning, Jr., *Success Is All That Was Expected: The South Atlantic Blockading Squadron During the Civil War* (Washington, D.C.: Brassey's, 2002), 1–4.
8. *Ibid.*, 5.
9. *Ibid.*, 8–13.
10. Browning, *Lincoln's Trident: The West Gulf Blockading Squadron During the Civil War* (Tuscaloosa, ALA: University of Alabama Press, 2015).
11. *Ibid.*, 76.
12. *Ibid.*, ix.

13. *Ibid.*
14. *Ibid.*, x.
15. *Ibid.*, 510–515.
16. *Ibid.*, 515.
17. Howard V. L. Bloomfield, *The Compact History of the United States Coast Guard* (New York: Hawthorn Books, 1966), 56.
18. *Ibid.*, 58.
19. *Ibid.*, 63.
20. Robert Scheina, "The Coast Guard at War," U.S. Department of Homeland Security and the United States Coast Guard, http://www.uscg.mil/history/articles/h_CGatwar.asp (accessed 23 October 2014).
21. Truman Strobridge, "The United States Coast Guard and the Civil War: The U.S. Revenue Marine, It's Cutters, and Semper Paratus," U.S. Department of Homeland Security and United States Coast Guard, http://www.uscg.mil/history/articles/Civil_War_Strobridge.asp (accessed November 12, 2014).
22. *Ibid.*
23. *Ibid.*
24. *Ibid.*

Chapter 12

1. Thomas P. Ostrom and John J. Galluzzo, *United States Coast Guard Leaders and Missions, 1790 to the Present* (Jefferson, NC: McFarland, 2015).
2. *Ibid.*, 1.
3. *Ibid.*, 10.
4. *Ibid.*, 14.
5. *Ibid.*, 56–58.
6. James M. McPherson, *Battle Cry of Freedom: The Civil War Era* (New York: Oxford University Press, 1988), 368, 392.
7. *Ibid.*, 369–370, 380–381.
8. *Ibid.*, 382.
9. *Ibid.*, 392–393, 417, 420, 426–427.
10. McPherson, *Ordeal by Fire: The Civil War and Reconstruction* (New York: McGraw Hill, 2001).
11. *Ibid.*, 206.
12. *Ibid.*
13. *Ibid.*, 255.
14. *Ibid.*, 192–195.
15. Dean Jobb, "The Cruise of the Tallahassee: The Confederacy's Last Great Raid on Union Shipping," *SEA History*, Summer 2015, 34–37.
16. *Ibid.*, 34.
17. *Ibid.*, 34–35.
18. *Ibid.*, 35.
19. *Ibid.*, 35–36.
20. *Ibid.*, 36–37.
21. *Ibid.*, 37.
22. William M. Fowler, Jr., *Under Two Flags: The American Navy in the Civil War* (Annapolis, MD: Blue Jacket Books, Naval Institute Press), 2001.
23. *Ibid.*, 38.
24. *Ibid.*, 42–44.
25. *Ibid.*, 300–303.
26. *Ibid.*, 300–309.
27. *Ibid.*, 309.
28. McPherson, *War on the Waters: The Union and Confederate Navies, 1861–1865* (Chapel Hill, NC: University of North Carolina Press), 2012.
29. *Ibid.*, 226.
30. *Ibid.*
31. *Ibid.*, 17–20.
32. *Ibid.*, 129.
33. Craig L. Symonds, *The Civil War at Sea* (Santa Barbara, CA: Praeger Publishers), 2009.
34. *Ibid.*, 14.
35. *Ibid.*, 17.
36. *Ibid.*, 62.
37. *Ibid.*, 67.
38. *Ibid.*, 37.
39. *Ibid.*, 163.
40. *Ibid.*, 170.

Chapter 13

1. Walter Star, *Steward: Lincoln's Indispensible Man* (New York: Simon & Schuster, 2012), 203–204.
2. Thomas P. Ostrom, and John J. Galluzzo, *United States Coast Guard Leaders and Missions, 1790 to the Present* (Jefferson, NC: McFarland, 2015), 52–53.
3. Stahr, *Seward*, 482–483.
4. *Ibid.*, 483–492.
5. Claus M. Naske and Herman E. Slotnick, *Alaska: A History of the 49th State* (Norman, OK: University of Oklahoma Press, 1987), 66–67, 277.
6. Ostrom and Galluzzo, *Coast Guard Leaders and Missions*, 11.
7. *Ibid.*, 54–55.
8. *Ibid.*, 57.
9. Truman R. Strobridge and Dennis L. Noble, *Alaska and the Revenue Cutter Service, 1867–1915* (Annapolis, MD: Naval Institute Press, 1999), 50–55, 122–126, 180–181.
10. *Ibid.*, 181.
11. Paul H. Johnson, "The Overland Expedition: A Coast Guard Triumph," http://www.uscg.mil/history/articles/johnson_overland_expedition… 1.

12. *Ibid.,* 2–4.

13. *Ibid.,* 4–6.

14. *Ibid.,* 6.

15. *Ibid.,* 7.

16. Irving H. King, *The Coast Guard Expands, 1865–1915* (Annapolis, MD: Naval Institute Press, 1996), 96–97.

17. *Ibid.,* 98–99.

18. Strobridge and Noble, *Alaska and the Revenue Cutter Service,* 128.

19. Stephen H. Evans, *The United States Coast Guard, 1790–1915: A Definitive History* (Annapolis, MD: The United States Naval Institute, 1949), 134–139.

20. *Ibid.,* 134–135.

21. *Ibid.,* 137–138.

22. *Ibid.,* 138.

23. *Ibid.,* 138–139.

24. *Ibid.,* 139.

25. Evans, *U.S. Coast Guard, 1790–1915,* 106.

26. *Ibid.,* 106–107.

27. *Ibid.,* 108–110.

28. *Ibid.,* 110–114.

29. *Ibid.,* 114.

30. *Ibid.,* 115–118.

31. *Ibid.,* 117.

32. *Ibid.,* 118–120.

33. *Ibid.,* 120.

34. *Ibid.*

35. King, *The Coast Guard Expands,* 36.

36. *Ibid.,* 52.

37. Strobridge and Noble, *Alaska and the Revenue Cutter Service,* 119.

38. *Ibid.,* 135–136.

39. *Ibid.,* 67–71.

40. Thomas J. Cutler, "'Admiral' Washington," *Proceedings,* October 2015, 92.

41. "Maritime Matters: Talk of Procuring Icebreakers Gains Traction in Washington," Megan Scully, Richard R. Burgess, and Otto Kreisher, SEAPOWER, October 2015, 6–7.

Chapter 14

1. Thomas P. Ostrom, and John J. Galluzzo, *United States Coast Guard Leaders and Missions, 1790 to the Present* (Jefferson, NC: McFarland, 2015), 11–12. 15, 58.

2. Stuart Murray, *Atlas of American Military History* (New York: Checkmark Books, 2005).

3. Craig L. Symonds, *Historical Atlas of the U.S. Navy* (Annapolis, MD: Naval Institute Press, 1995), 110.

4. Robert Pendleton and Patrick McSherry, "The U.S. Revenue Cutter Service in the Spanish American War," http://www.spanamwar.com/USRCS.htm, 1–17.

5. *Ibid.,* 2–3.

6. *Ibid.,* 3.

7. *Ibid.*

8. *Ibid.*

9. *Ibid.,* 4.

10. *Ibid.,* 4–5.

11. *Ibid.,* 6–7.

12. *Ibid.,* 7–8.

13. *Ibid.,* 8–9.

14. *Ibid.,* 9.

15. *Ibid.,* 9–10.

16. *Ibid.,* 10–12.

17. Thomas P. Ostrom, and John J. Galluzzo, *United States Coast Guard Leaders and Missions, 1790 to the Present* (Jefferson, NC: McFarland, 2015), 58.

18. Pendleton and McSherry, *The USRCS in the Spanish-American War,* 10–12.

19. *Ibid.,* 11–12.

20. *Ibid.,* 12.

21. Ostrom and Galluzzo, *United States Coast Guard Leaders and Missions,* 74, 182.

22. Pendleton and McSherry, 12–13.

23. Alex R. Larzelere, *The Coast Guard in World War I* (Annapolis, MD: Naval Institute Press, 2003), 180–181.

24. Pendleton and McSherry, 13.

25. *Ibid.,* 14–16.

26. Irving H. King, *The Coast Guard Expands, 1865–1915: New Roles, New Frontiers,* Annapolis, MD: Naval Institute Press, 1996), 109–110, 122.

27. Stephen H. Evans, *The United States Coast Guard, 1790–1915: A Definitive History* (Annapolis, MD: The United States Naval Institute, 1949.

28. *Ibid.,* 160.

29. *Ibid.,* 161–162.

30. *Ibid.,* 163–164.

31. *Ibid.,* 164–165.

32. *Ibid.*

33. *Ibid.,* 165.

34. *Ibid.,* 166.

35. *Ibid.*

36. *Ibid.,* 167, 168.

37. *Ibid.,* 169.

38. *Ibid.,* 171–172.

39. *Ibid.,* 172.

40. *Ibid.*

41. *Ibid.,* 172–175.

42. *Ibid.,* 175.

43. *Ibid.,* 175–176.

44. John Whiteclay Chambers II (Editor-in-Chief), *The Oxford Companion to American*

Military History (New York: Oxford University Press, 1999, 145.

45. Ibid., 488–489.

Chapter 15

1. "Commodore Ellsworth P. Berthold." http://www.uscg.mil/pacarea/cgcBertholf/history/ComdtBio.asp (accessed 03 December 2015).

2. Ibid.

3. Ibid.

4. John Whiteclay Chambers II (Editor-in-Chief), The Oxford Companion to American Military History (New York: Oxford University Press, 1999), 145.

5. Ibid.

6. Craig L. Symonds, The Naval Historical Atlas of the U.S. Navy (Annapolis, MD: Naval Institute Press, 1995), 120–121. Cartography by William J. Clipson.

7. Ibid., 130–131.

8. Stuart Murray, Atlas of American Military History (New York: Checkmark Books, 2005), 136–137.

9. Ibid.

10. Irving H. King, The Coast Guard Expands: New Roles, New Frontiers, 1865–1915 (Annapolis, MD: Naval Institute Press, 1996), 243.

11. Ibid.

12. Thomas P. Ostrom and John J. Galluzzo, United States Coast Guard Leaders and Missions, 1790 to the Present (Jefferson, NC: McFarland, 2015), 74.

13. Ibid., 74–75.

14. Ibid., 79.

15. Ibid., 84.

16. Ibid., 84–85.

17. Ibid., 96.

18. Ibid., 157, 172.

19. Thomas P. Ostrom, The United States Coast Guard and National Defense (Jefferson, NC: McFarland, 2012), 14.

20. Alex R. Larzelere, The Coast Guard in World War I: The Untold Story (Annapolis, MD: Naval Institute Press, 2003), 99.

21. Ibid., 104–105.

22. Ibid., 106.

23. Ibid., 171.

24. Ibid., 173–179.

25. Ibid., 179.

26. Ibid.

27. Ibid., 179–180.

28. Ibid., 180.

29. Ibid.

30. Ibid., 185.

31. Ibid., 237–240.

32. Ibid., 241.

33. Ibid., 240–255.

Chapter 16

1. Robert Scheina, "The U.S. Coast Guard at War: A History," U.S. Department of Homeland Security, and United States Coast Guard, http://www.uscg.mil/history/articles/h_CGatwar.asp (accessed 08 January 2015), 1–2.

2. Ibid., 2.

3. Ibid., 2–4.

4. Donald L. Canney, The Confederate Steam Navy, 1861–1865 (Atglen, PA: Schiffer Publishing Ltd.), 12015, 5, 140–141.

5. Ibid., 5.

6. Alan Fraser Houston, "The Many Faces of Lady Sterling," Naval History, April 2016, 50–55.

7. Ibid., 5–6.

8. Ibid., 6–7.

9. Ibid., 7–8.

10. "Charles Satterlee," Wikipedia Encyclopedia, http://en.wikipedia.lrg/wiki/Charles_Satterlee (accessed 07 February 2016).

11. "Tampa: Revenue Cutter," NiSource Online: Section Patrol Craft Photo Archive, USCGC Tampa, ex–USCGC Miami, ex–USRC Miami, http://www.navsource.org/archives/12/179894.htm (accessed 07 February 2016). Information on the NavSource Online website provided by Joseph M. Radigan, copyright 1996–2016 NavSource Naval History.

12. "Together We Served," The Military Honor Wall, Captain Charles A, Satterlee, https://coastguard.togetherweserved.com/uscg/servlet/tws.weba... (accessed 07 February 2016). Copyright Togetherweserved.com Inc. 2003–2011.

13. Scheina, "The U.S. Coast Guard at War," 8.

14. Ibid.

15. Ibid.

16. Ibid., 9.

17. William H. Thiesen, "The Coast Guard's Aerial Visionary," Naval History, April 2016, 36–41.

18. Ibid., 39.

Epilogue

1. David O. Whitten, "Review of The U.S. Navy: A Concise History," by Craig L. Symonds (2015), Sea History (Summer 2016), 63.

2. Peter Stanford, "The Cape Horn Road, Part II: How the Sails of the Square-rigged Ship Got Their Names (2000)," *Sea History* (Summer 2016), 18–21.

3. Peter H. Daly, "An Independence Day Message from the Independent Forum," U.S. Naval Institute, newsletters@usni.org, July 4, 2016, 1–2 (accessed Monday July 4, 2016).

4. *Ibid.*

5. Tom Beard, Editor-in-Chief, *The Coast Guard* (Seattle, WA: Foundation for Coast Guard History, and Hugh Lauter Levin Associates, 2004), 205.

6. *Ibid.*

7. *Ibid.*, 206–207.

Bibliography

Ayer, Robert, editor and Gary M. Thomas, executive director, "Mexican War: 1846–1848." *The Cutter.* "Mexican War: 1846–1848." *The Cutter,* Summer/Fall 2014, 20.

Beard, Tom, Jose Hanson, and Paul C. Scotti, editors. *The Coast Guard.* Seattle, WA: Foundation for Coast Guard History/Westbrook, CT: Hugh Lauter Levin Associates, 2004.

Billington, Ray Allen, and Martin Ridge. *Westward Expansion: A History of the American Frontier.* New York: Macmillan Publishing Co., Inc., 1982.

Bloomfield, Howard V. L. *The Compact History of the United States Coast Guard.* New York: Hawthorn Books, Inc., 1966.

Borneman, Walter R. *1812: The War That Forged A Nation.* New York: HarperCollins Publishers, 2005.

Browning, Robert M., Jr. *Lincoln's Trident: The West Gulf Blockading Squadron During the Civil War.* Tuscaloosa, Alabama: University of Alabama Press, 2015.

_____. *Success Is All That Was Expected: The South Atlantic Blockading Squadron During the Civil War.* Washington, D.C.: Brassey's, Inc., 2002.

Canney, Donald L. *The Confederate Steam Navy.* Atglen, PA: Schiffer Publishing, Ltd., 2015.

_____. *U.S. Coast Guard and Revenue Cutters, 1790–1935.* Annapolis, MD: Naval Institute Press, 1995.

Catton, Bruce. *The Civil War.* Boston, MA: Houghton Mifflin Co., First Mariner Books, 2004.

Chaffin, Tom. *Sea of Gray: The Around The World Odyssey of the Confederate Raider Shenandoah.* New York: Hill and Wang, 2006.

Chambers, John Whiteclay II, et al. *The Oxford Companion to American Military History.* New York: Oxford University Press, 1999.

Chapelle, Howard I. *The History of American Sailing Ships.* New York: W.W. Norton and Company, Inc., 1935.

"Charles Satterlee." *Wikipedia Encyclopedia.* Accessed February 7, 2016. https://en.wikipedia.org/wiki/Charles_Satterlee.

"Commodore Ellsworth P. Bertholf." *United States Coast Guard Pacific Area.* Accessed December 3, 2015. http://www.uscg.mil/pacarea/cgcBertholf/history/.

Cutler, Thomas J. "'Admiral' Washington." *Proceedings,* October 2015.

Daly, Peter H. "An Independence Day Message from the Independent Forum." *U.S. Naval Institute.* July 4, 2016. Accessed July 4, 2016. 1–2.

Dougherty, Kevin J. *Ships of the Civil War, 1861–1865.* New York: Metro Books/Sterling Publishing Co., Inc., Amber Books, Ltd., 2013.

"Eighteenth, Nineteenth & Early Twentieth Century Revenue Cutters." U.S. Department of Homeland Security and United States Coast Guard. U.S. Coast Guard Historian's Office. Accessed October 29, 2014. http://www.uscg.mil/history/webcutters/USRC_Photo_Index.asp.

"Ellsworth P. Bertholf." U.S. Department of Homeland Security and United States Coast Guard. U.S. Coast Guard Historian's Office. October 25, 2014. http://www.uscg.mil/history/people/EPBertholfBio.asp.

Evans, Stephen H. *The United States Coast Guard, 1790–1915: A Definitive History.* Annapolis, MD: The United States Naval Institute, 1949.

Faust, Patricia L., ed., *Historical Times Illustrated Encyclopedia of the Civil War.* New York: Harper & Row, Publishers, 1986.

"First Ten Cutters." U.S. Department of Homeland Security and United States Coast Guard. U.S. Coast Guard Historian's Office. Accessed October 29, 2014. http://www/uscg.mil/history/webcutters/First_Cutters.asp.

Fowler, William M., Jr. *Under Two Flags: The American Navy in the Civil War.* Annapolis, MD: Blue Jacket Books, 2001.

Galluzzo, John J. "1790–1915: 'That Others Might Live.'" In *The Coast* Guard, edited by Tom Beard, Seattle, WA: Foundation for Coast Guard History, and Hugh Lauter Levin Associates, 2004.

"Great Lakes Revenue Vessels." *United States Coast Guard and the U.S. Department of Homeland Security.* U.S. Coast Guard History Program. Accessed December 17, 2014.

Hansen, Harry. *The Civil War: A History.* New York: Signet Classics and the Penguin Group, 2002.

Houston, Alan Fraser. "The Many Faces of *Lady Sterling.*" *Naval History,* April 2016.

Hughes, Dwight S. "Pirate, Privateer, or Man-of-War." *Naval History,* February 2016.

Jobb, Dean. "The Cruise of the Tallahassee: The Confederacy's Last Great Raid on Union Shipping." *Sea History,* Summer 2015.

Johnson, Paul H. "The Overland Expedition: A Coast Guard Triumph." Accessed July 1, 2015.
_____. "The Search for Captain Frederick Lee." *The Bulletin,* March/April, 1977. http://www.uscg.mil/history/articles/johnson_overland.expedition.

King, Irving H. *The Coast Guard Expands, 1815–1915.* Annapolis, MD: Naval Institute Press, 1996.

Krom, Richard G. *The First Minnesota: Second to* None. Rochester, MN: Richard G. Krom, self-published; Brainerd, MN: Bang Printing, 2010.

Lardas, Mark. *CSS Alabama vs USS Kearsarge: Cherbourg, 1864.* Oxford, United Kingdom, and Long Island City, New York: Osprey Publishing, Ltd., 2011.
_____. *Great Lakes Warships, 1812–1815.* Long Island City, NY: Osprey Publishing, Ltd., 2012.

Larzelere, Alex R. *The Coast Guard at War: Vietnam, 1965–1975.* Annapolis, MD: Naval Institute Press, 1997.
_____. *The Coast Guard in World War I.* Annapolis, MD: Naval Institute Press, 2003.

"Loss of the U.S. Revenue Schooner *Gallatin,* 1 April 1813 in Charleston Harbor, South Carolina." *United States Coast Guard, Department of Homeland Security. U.S. Coast Guard History Program.* Accessed November 16, 2014.

"(Revenue Cutter) *Louisiana, 1804.*" United States Coast Guard, Coast Guard Historian's Office. Accessed December 20, 2014. https://www.uscg.mil/history/…Louisiana.pdf.

McPherson, James M. *Battle Cry of Freedom: The Civil War Era.* New York: Oxford University Press, 1988.
_____. *Embattled Rebel: Jefferson Davis as Commander in Chief.* New York: Penguin Press, 2014.
_____. *Ordeal by Fire: The Civil War and Reconstruction.* New York: McGraw-Hill, 2001.
_____. *War on the Waters: The Union and Confederate Navies, 1861–1865.* Chapel Hill: University of North Carolina Press, 2012.

Morison, Samuel Eliot, Henry Steele Commager, and William Leuchtenburg. *The Growth of the American Republic.* vol. I. New York: Oxford University Press, 1969.

Morison, Samuel Eliot. *The Oxford History of the American People.* New York: Oxford University Press, 1965.

Murray, Stuart. *Atlas of America Military History.* New York: Checkmark Books, 2005.

Naske, Claus M., and Herman E. Slotnick. *Alaska: A History of the 49th State.* Norman, Oklahoma: University of Oklahoma Press, 1987.

Noble, Dennis L. *That Others Might Live: The U.S. Life-Saving Service, 1878–1915.* Annapolis, MD: Naval Institute Press, 1994.

Ostrom, Thomas P. *The United States Coast Guard and National Defense.* Jefferson, NC: McFarland, 2012.
_____. *The United States Coast Guard, 1790 to the Present.* Oakland, OR: Elderberry/Red Anvil Press, 2006.
_____, and John J. Galluzzo. *United States Coast Guard Leaders and Missions, 1790 to the Present.* Jefferson, NC: McFarland, 2015.

Patterson, Benton Rain. *The Mississippi River Campaign, 1861–1863: The Struggle for Control of the Western Waters.* Jefferson, NC: McFarland, 2010.

Pendleton, Robert, and Patrick McSherry. "The U.S. Revenue Cutter Service in the Spanish American War." Accessed October 18, 2015. http://www.spanamwar.com/USRCS.htm.

Scheina, Robert L. "The Coast Guard at War." Accessed October 23, 2014. http://www.uscg.mil/history/articles/hCGatwar.asp.

———. "The Coast Guard at War." Accessed January 8, 2015. http://www.uscg.mil/history/articles/h_CGatwar.asp.

Schweikart, Larry, and Michael Allen. *A Patriot's History of the United States*. New York: Sentinel/Penguin Group, 2007.

Scully, Megan, Richard R. Burgess, and Otto Kreisher. "Maritime Matters: Talk of Procuring Icebreakers Gains Traction in Washington." *SEAPOWER*, October 2015.

Stahr, Walter. *Seward: Lincoln's Indispensable Man*. New York: Simon & Schuster, 2012.

Stampp, Kenneth M., ed. *The Causes of the Civil War*. New York: Simon & Schuster, Inc., 1991.

Stanford, Peter. "The Cape Horn Road, Part II: How the Sails of the Square-rigged Ship Got Their Names (2000)." *Sea History*, Summer 2016.

Strobridge, Truman. "The United States Coast Guard and the Civil War: The U.S. Revenue Marine, Its Cutters, and Semper Paratus." *U.S. Department of Homeland Security and the United States Coast Guard*. Accessed November 12, 2014. http://www.uscg.mil/history/articles/Civil_War_Strobridge.asp.

Strobridge, Truman R., and Dennis L. Noble. *Alaska and the U.S. Revenue Cutter Service, 1867–1915*. Annapolis, MD: Naval Institute Press.

Symonds, Craig L. *The Civil War at Sea*. Santa Barbara, CA: Praeger Publishers, 2009.

———. *Decision at Sea: Five Naval Battles That Shaped American History*. Oxford, NY: Oxford University Press, 2005.

———. *Historical Atlas of the U.S. Navy*. Annapolis, MD: Naval Institute Press, 1995.

———. *The U.S. Navy: A Concise History*. New York: Oxford University Press, 2016.

"Tampa: Revenue Cutter." *NavSource Online: Section Patrol Craft Photo Archive*, USCGC Tampa, *ex*-USCGC Miami, *ex*-USRC Miami. Accessed February 7, 2016. http://www.navsource.org/archives/12/179894.htm.

Thiesen, William H. "Benjamin Hiller & the Cutter *Pickering* in the Quasi-War with France." *SEA HISTORY*. Spring 2008.

———. "Captain Frederick Lee and Cutter Eagle in the War of 1812." *The Bulletin*. U.S. Coast Guard Academy Alumni Association, June 2012.

———. "The Coast Guard's Aerial Visionary." *Naval History*. April 2016.

———. "They Did Their Duty as Became American Sailors." *Naval History*. August 2014.

———. "United States Coast Guard, War of 1812." Revenue Cutter Operations and the Core Coast Guard Missions." U.S. Coast Guard Historian's Office, U.S. Coast Guard Headquarters, Washington, D.C. Accessed January 4, 2015. 1–15.

———. "With a 'Spirit and Promptitude.'" *Proceedings*. August 2012.

Tindall, George Brown, and David E. Shi. *America: A Narrative History*. New York: W. W. Norton & Company, 1996.

"Together We Served." *The Military Honor Wall. Captain Charles A. Satterlee*. Accessed February 7, 2016. https://coastguard.togetherweserved.com/uscg/servlet/tws.weba.

Turner, Thomas R. *The Civil War: 101 Things You Didn't Know About It*. Avon, MA: Adams Media, 2007.

"U.S. Coast Guard: Historical Overview." U.S. Department of Homeland Security and United States Coast Guard. Coast Guard Historian's Office. Accessed December 20, 2014. http://www.uscg.mil/history/articles/h_USCGhistory.asp.

"War of 1812 Revenue Cutter and Naval Operations." United States Coast Guard and U.S. Department of Homeland Security. U.S. Coast Guard History Program. U.S. Coast Guard Historian's Office. Accessed December 15, 2014.

"When the U.S. Invaded Canada." *The Week*, August 10, 2012.

Whitten, David O. "Review of *The U.S. Navy: A Concise History*, by Craig Symonds (2015). *Sea History*, Summer 2016, 18–21.

Index

Numbers in **bold italics** indicate pages with illustrations